ACOUSTIC COLONIALISM

DISSIDENT ACTS

A series edited by Macarena Gómez-Barris and Diana Taylor

ACOUSTIC COLONIALISM

ACTS OF MAPUCHE INTERFERENCE

LUIS E. CÁRCAMO-HUECHANTE

DUKE UNIVERSITY PRESS *Durham and London* 2025

Printed and bound by CPI Group (UK) Ltd, Croydon, CR0 4YY

Project Editor: Michael Trudeau
Designed by Matthew Tauch
Typeset in Warnock Pro by Westchester Publishing
Services

Library of Congress Cataloging-in-Publication Data
Names: Cárcamo-Huechante, Luis E. author
Title: Acoustic colonialism : acts of Mapuche interference / Luis E. Cárcamo-Huechante.
Other titles: Dissident acts
Description: Durham : Duke University Press, 2025. | Series: Dissident acts | Includes bibliographical references and index.
Identifiers: LCCN 2025012355 (print)
LCCN 2025012356 (ebook)
ISBN 9781478032632 paperback
ISBN 9781478029298 hardcover
ISBN 9781478061533 ebook
Subjects: LCSH: Mapuche Indians—Chile—Social life and customs | Mapuche Indians—Chile—Folklore | Mapuche Indians—Chile—Music—History and criticism | Sound—Social aspects—Chile—History—19th century | Indigenous peoples—Chile—Ethnic identity | Cultural appropriation—Chile | Oral tradition—Chile | Settler colonialism—Chile—History
Classification: LCC F3126 .C3525 2025 (print) | LCC F3126 (ebook) | DDC 305.898/72083—dc23/eng/20250416
LC record available at https://lccn.loc.gov/2025012355
LC ebook record available at https://lccn.loc.gov/2025012356

Cover art: Illustration by Matthew Tauch.

To María Clara Lefno Huechante (1905–95)

To Jovita Huechante Lefno (1941–2023)

To the Mapuche People

CONTENTS

AUTHOR'S NOTE

Alphabets for Mapudungun

The first attempts in Chile to transcribe Mapudungun, the language of the Mapuche People, into writing occurred in the late nineteenth and early twentieth centuries, mostly at the hands of Catholic missionaries who happened to have some training or ambitions as linguists (i.e., Félix José de Augusta, Ernesto Wilhelm de Moesbach). Subsequent attempts throughout the twentieth century also deployed non-Indigenous researchers of Mapudungun, who used Mapuche persons and communities as mere "informants." By the early 2000s, when Mapuche scholars started to play an important role in transliterating their language, there were already more than twenty alphabets or writing grammars for Mapudungun. By the 1990s, the most widely legitimized alphabets among the Mapuche intellectual and political community were the Unified Alphabet and Raguileo Alphabet. Today the Azümchefe Alphabet is also in wide use due to its creation by the National Corporation for Indigenous Development (CONADI, by its acronym in Spanish), the Indigenous affairs office of the State of Chile. In this book, I use the Unified Alphabet due to its accessibility for non–speakers of Mapudungun and the Raguileo Alphabet in cases such as the Mapuche radio show discussed in chapter 4, which had adopted it as their official writing grammar. In other cases where the sources differ from both I stick with the primary sources' alphabets.

UNIFIED ALPHABET · This alphabet was drafted and approved by the scholarly Sociedad Chilena de Lingüística (SOCHIL) at a congress that took place in May 1986 at the Catholic University of Chile, Temuco. The majority consensus on it held that as a writing grammar it would facilitate the process of language learning and would be more legible for Mapuche and non-Mapuche students of Mapudungun. Mapuche scholars, such as María Catrileo, participated in these deliberations and endorsed the use of this alphabet.

RAGUILEO ALPHABET · This writing grammar of Mapudungun was created by independent Mapuche scholar Anselmo Raguileo in 1982. Without giving up the use of the Latin letter and graphic system, his writing grammar favors the generation of graphemes that aim to differentially follow the phonetics of Mapuzugun. Many organizations of the Mapuche movement valued this writing grammar as a political effort to prioritize a practice of Mapuche autonomy in the terrain of language.

TABLE FM.1

UNIFIED ALPHABET	RAGUILEO ALPHABET	EXAMPLES IN KEY WORDS		
A	A			
CH	C	Mapuche	Mapuce	People of the Land, or People of the Earth
D	Z	Mapudungun	Mapuzugun	Language of the Mapuche/Mapuce
E	E			
F	F			
NG	G	Ngulu Mapu	Gulu Mapu	The Lands of the West (what is today central and southern Chile)
I	I			
K	K			

UNIFIED ALPHABET	RAGUILEO ALPHABET	EXAMPLES IN KEY WORDS		
L	L			
L̲	B			
LL	J	Wallmapu	Wajmapu	The whole Mapuche ancestral territory, including areas of central and southern Argentina and central and southern Chile
M	M			
N	N			
N	H			
Ñ	Ñ			
G	Q			
O	O			
P	P			
R	R			
S	S			
T	T			
T				
TR	X	Nütram	Nvxam	Conversation, talk
U	U			
Ü	V	Ül	Vl	Mapuche chant
W	W			
Y	Y			
SH				

Table FM.1 is a comparative table of the graphemes across these two alphabets along with examples of graphemic variations for some of the keywords in Mapudungun/Mapuzugun used throughout this book.

ON TRANSLATIONS AND TRANSCRIPTIONS

For transcriptions and some of the translations from Mapudungun/ Mapuzugun, I have had the generous collaboration of the following Mapuche fellows: Elisa Avendaño Curaqueo, Margarita Elizabeth Huenchual Millaqueo, Claudia Ingles Hueche, Pablo Andrés Millalen Lepin, and José Alfredo Paillal Huechuqueo.

All translations into English from other languages are mine, unless translation is attributed to others. Besides the assistance of my editors, Anitra Grisales, Nancy Sorkin, and Robert Weiss, for some more translations, Joseph Pierce (Cherokee) has provided important help.

MAÑUMTUN / ACKNOWLEDGMENTS

Mañumfin ta chi Mapu ñi kimün ka ñi newen. Fentren mañum ta iñ Mapunche kuifikeche ka. Chi Mapu ñi newen kellui ta ñi lifru—küdaw meu. Kellufi ta ñi rakiduam ka ta ñi piuke ñi dewmaiafiel tüfachi wirin.

Fentren mañum ta ñi Huechante Lefno reñma, rumel ta ñi püllü ka ta ñi piuke mew. Ta ñi chuchu ("mami") María Clara Lefno Huechante yeniefin ta ñi püllü mew. Rumel ta ñi ñuke Jovita Huechante Lefno ñi newen. Mañumün ka ta ñi Huechante pu weku, pu ñukentu, ka pu müna. Mañumün ta ñi reñma ka ta iñ küpalme ñi küme üy: Weche Antü.

Dewmanentuam tüfachi lifru mülei ta küme kellu Mapuche, pu peñi ka pu lamngen: Francisco Bascur, Margarita Elizabeth Huenchual Millaqueo, Claudia Ingles Hueche, Pablo Millalen Lepin ka José Paillal Huechuqueo. Ka ta ñi fentren mañum ta chi pu lamngen ka chi pu peñi Clara Antinao Varas, Elisa Avendaño Curaqueo, Ramón Curivil Paillavil, Graciela Huinao, Leonel Lienlaf, Juan Atanacio Meulen Trayanado (1945–2025), Ana Millaleo, Paul Paillafilu ka Elías Paillan Coñoepan.

Ka mañumün ta mün nütramkan ka ta mün newen fentren Mapuche wenüy: Margarita Calfío Montalva, Seba Calfuqueo, Jacqueline Curaqueo, Herson Huinca Piutrin, Cristina Llanquileo, Pablo Mariman Quemenado, Roxana Miranda Rupailaf, Héctor Nahuelpan ka Moira Millán.

Mañumün kom pu peñi, pu lamngen ka pu wenüy. Mañumün ta mün küñiwtun meu, ta mün poyen meu, ka ta mün newen meu.

....

Besides expressing gratitude to the Mapu and the Mapuche as I have expressed it in Mapudungun above, as an Indigenous guest in Turtle Island—one of the ancestral names for what is now called "North America"—it is important for me to acknowledge its lands and its peoples who have hosted me here. In central Texas, where Native life has been subject to a hegemonic history of colonial erasure and elimination, I would like to acknowledge these lands as territories of life and movement of several tribes across centuries, such as the Apache, Caddo, Cherokee, Coahuiltecan, Comanche, Karankawa, Lipan, Seminole, and Tonkawa. My knowledge has been enriched and expanded by the conversations that I have had with Native community members in the city of Austin and in central Texas. In this respect, I want to express my deep gratitude to elder Marika Alvarado (Lipan Mescalero), a Medicine Woman—or what we the Mapuche call *Lawentuchefe*—who has always been there for me, to offer guidance and medicinal care. I also want to acknowledge Dr. Mario Garza and Maria Rocha (Miakan-Garza Band), Marina Islas (Lipan Apache), Nan Blassingame (Cheyenne and Arapaho) and Robert Bass (Great Promise for American Indians, Austin), who have also been important hosts for me in the lands of what is today central Texas.

When I started to conceive this book, the feedback I received from a stellar group of scholars during a 2013–14 residential fellowship at the National Humanities Center was crucial. There I was able to particularly enjoy enriching dialogues about my research with colleagues Charles McGovern, James Maffie, Julie Greene, Harvey J. Graff, Orin Starn, and Katya Wesolowski. At the University of Texas at Austin, the support of the College of Liberal Arts and dialogues with colleagues and students at the Program in Native American and Indigenous (NAIS), the Lozano Long Latin American Studies Institute (LLILAS), and the Department of Spanish and Portuguese played an important role in this process. I also would like to acknowledge the colleagues who invited me to their home institutions in the United States, Canada, Brazil, Argentina, Chile, Cambridge (UK), and Scotland to present samples of my work on colonialism, sound, and indigeneity.

Friends and close colleagues were also critical to the formulation and development of this book. Emil' Keme, Gwen Kirkpatrick, Ana María Ochoa Gautier, Francine Masiello, Joseph Pierce, and Richard Shain offered me attentive reading and brilliant feedback on earlier versions of different chapters or sections of the manuscript. Conversations and collaborations with Luis Gamboa, Jorge Montealegre, and Juan Yilorm Martínez were of great relevance in accessing archival sources in Chile.

While I was outlining, drafting, and writing *Acoustic Colonialism*, Sam Vong's beautiful intellectual and affectionate companionship was tremendously important, leaving an indelible trace in me and on this book. I also would like to thank several friends in a special way, for just supporting me during the years of writing. My heartfelt gratitude for being part of my life, Rosita Celedón, Héctor Domínguez-Ruvalcaba, Brad Epps, Arvind Gopinath, José Antonio Mazzotti (1961–2024), Jorge Pérez, José Salomon, Ximena Sgombich, Shannon Speed, Anand Bala Subramaniam, Juan Pablo Sutherland, Marcy Schwartz, Enzo Vasquez Toral, and Baruk Villar. Also, for their continuous encouragement and support, my enormous gratitude to Jossianna Arroyo-Martínez, Jason Borge, Natalia Brizuela, Debra Castillo, Karma Chavez, Barbara Corbett, Ann Cvetkovich, Laura Demaría, Leila Gómez, J. Kēhaulani Kauanui, Lorraine Leu, Kelly McDonough, Lynn Moore, Carlos Ramos, Sonia Roncador, Ruth Rubio-Gilbertson, Pauline Strong, and Anthony Webster. Additionally, my nine years as director of the Program in Native American and Indigenous Studies (NAIS) at UT Austin offered me a stimulating milieu to undertake this writing labor. Collaboration with colleagues Jennifer Graber and Luis Urrieta as well as the assistance of Anna Gallio constituted a supportive environment.

For the actual writing of *Acoustic Colonialism*, I do not have enough words to express my immense gratitude to Morgan Blue, Anitra Grisales, Nancy Sorkin, and Robert Weis, who helped me as editors of my manuscript at several stages, for a period of almost two years. Anitra, Nancy, and Robert are omnipresent spirits in every line of this book. Thank you!

Finally, special thanks to the marvelous colleague and scholar Macarena Gómez-Barris, who was always there for *Acoustic Colonialism*, enthusiastically welcoming me to the Dissident Acts series she codirects with Diana Taylor at Duke University Press. Duke UP editors Ken Wissoker and Kate Mullen have been ardent proponents of its publication, offering me effective and generous guidance at every step of the editorial process. Substantive feedback from anonymous readers enriched and improved the manuscript in style, form, and content. I extend my gratitude to the professional staff of Duke University Press who worked on multiple aspects of what is now *Acoustic Colonialism*.

Introduction

Tralcao *is a word with strong resonances. It comes from* tralkan, *which, in Mapudungun, the language of the Mapuche People, means "thunderstorm." Located within what is today known as "southern Chile," Tralcao is the name of the rural community where I grew up and spent my childhood and part of my adolescence before moving to the nearby city of Valdivia in the mid-1970s to attend secondary school. At the time, Tralcao was a Mapuche-Williche community, with few non-Indigenous families. Williche—"People of the South" or "People from the South"—is a branch of the Mapuche (People of the Earth). As was the case in many Williche territories, Catholic evangelization colonized the minds of many community members over centuries. In a long conversation, elder Arturo Navarro Caurepan told me the story of Christian colonization and destruction of what used to be a* Lof *(Mapuche local territorial unity; interview by author, 2007). By the late nineteenth century, there was a collective fear among the Williche in our region of a potential Chilean military invasion and forced removal, similar to what had occurred to the north in the so-called Pacification Campaign of the Araucanía. Because of this threatening environment, the* longko *(community head) from our Huechante kinship requested protection from the Capuchin mission that had settled in Mariquina, an area located some fifteen and a half miles north of Tralcao. The Capuchin priests demanded that the Mapuche abandon "witchcraft" (Mapuche ceremonies), speak only "the language of God" (Spanish), and cede most of the community lands to build the Mission of Pelchuquin. Under these conditions, the whole Lof was dismembered and relocated to what is today Tralcao, undergoing a drastic land reduction along with spiritual, linguistic, and cultural subjugation. Our phonic and linguistic environment was severely altered. As a term rooted in Mapudungun but transmuted through the phonetics of Spanish, the very name Tralcao is simply a linguistic disfigurement of what originally could have been Tralkawe, "place of thunderstorm." The name honored the very nature of the land where the Huechante, the Pangui, the Lefno, and other Williche kinships dwelled. The twinkling of lightning and the booming of thunder were common during the long, rainy winters of Tralcao, and they remain in my mind as markers of a natural environment sonically alive. Under the effect of the thunderstorms, everything seemed to tremble in Tralcao. Our human voices became tenuous, our portable radio got stuck in a series of screeches. Despite Chilean and Christian colonization, the sounds that prevailed in our environment were always those of the Mapu (land, earth, universe): the thunder, the rain, the wind. For us, they were the enduring sounds of the Füta Willi Mapu—the Great Lands of the South.*

....

The phonetic Hispanicization of the very name "Tralcao" is a telling example of what I conceptualize as *acoustic colonialism*. Valdivia, the name of the main settler city near Tralcao, also resonates with colonial history. The Spanish conqueror Pedro de Valdivia (1497–1553) founded the city in 1552 as one of the first urban settlements in Mapuche territories.[1] By the late eighteenth century, Jesuit missionaries founded the Mission of San José in the area the Mapuche knew as *Mariküga*—today, Mariquina—located north of Valdivia and not far from Tralcao.[2] The establishment of this mission aimed to consolidate the enterprise of Christian evangelization that had been set in motion during the previous centuries of Spanish colonial rule.[3] After the establishment of the Chilean nation-state in 1810, this colonial process would continue through the expansion of Chilean modes of government at the expense of Indigenous lifeways. In 1826, the Valdivia Province was officially instituted and became a privileged place for colonizers to settle, specifically on Williche lands as well as over the coastal territories of the Lafkenche (People of the Coast). By the mid-nineteenth century, occupying Williche lands became an important goal for German settlers, whom Chilean "national" and "provincial" authorities had recruited and sponsored to strengthen their plans of colonization. This process gave settlers access to Native lands, leading to the formation of large country estates (fundos) in Mapuche-Williche territories. This settler private-property expansion entailed the removal and displacement of Indigenous communities, usually based on the transfer of lands under "fiscal" jurisdiction.

As this Chilean state-sponsored project of land and culture dispossession gained ground in the Mapuche-Williche and Mapuche-Lafkenche territories Catholic missionaries joined this process. In the 1840s, a new group of Catholic missionaries, the Capuchin, arrived in the region of Mariquina.[4] After consolidating the Mission of San José de la Mariquina in the 1860s, the Capuchins decided to launch the Mission of Pelchuquin on lands that had historically been under the care of the Huechante kinship. Our ancestral territory was dramatically affected by this colonizing process. Our Williche community ended up reduced to what is currently Tralcao.

Being part of a colonized community was an everyday experience during my childhood and early adolescence in Tralcao. It exposed me to the omnipresence of acoustic colonialism resulting from the linguistic, religious, cultural, and geopolitical absorption of our Williche rural community life into the framework of Chilean society. In places like Tralcao, Catholic evangelization by the Capuchin missions had either banned or silenced Mapuche ceremonial and religious practices from public spaces. What Jessica Bissett Perea (Dena'ina) calls "missionary colonialism" (2021) was embedded in the everyday life of our Williche aural and sonic experience.[5] By the 1960s, the sounds of Christian preaching, prayers, and songs dominated Tralcao. The elementary school, a state institution where my studies began, was in the community's small Catholic chapel. Linguistically and aurally, our community experience was overdetermined by the erasure of Mapudungun and the imposition and naturalization of Spanish. At the school, the compulsory secular rite of singing and listening to the Chilean national anthem every Monday was also critical to the Chilean colonization of our ears as Williche children.

We also became accustomed to hearing the accents of German and other non-Indigenous settlers who owned and managed the country estates that made a colonizing ring around the Tralcao community. This local coexistence between state and religious institutions, along with powerful settler landowners, is illustrative of the alliance that supported Chilean colonialism in most of the Mapuche-Williche territories.

THE WAVES OF ACOUSTIC COLONIALISM

This book emerges from this Mapuche-Williche story as a personal and communal experience entangled in a colonized setting. My writing is a critical effort to liberate my own subject position from these contradictory colonial knots and moorings. The effort has entailed my involvement in a wide range of Mapuche engagements, from studying Mapudungun to weaving forms of politico-intellectual and cultural initiatives to support the Mapuche land-back movement. This book embodies and embraces a will to be in tune with the voices and sounds that are currently resurging as part of the historical struggle of Ngulu Mapu and, more extensively, Wallmapu—the broader Mapuche territory.

Underlying this personal and collective experience is the concept of colonialism, which I understand as the occupation of an Indigenous territory by the invasive presence and actions of exogenous bodies, agents, forces, relations, structures, and infrastructures. Acoustic and sonic dimensions of colonialism are embedded and interwoven in its material, discursive, and symbolic deployment. At the intersection of colonizing and colonized societies, the realm of sound always becomes an arena of movement—of fluidity as well as of tension and clash. Neither neutral nor isolated, sound is a relational phenomenon entangled in webs of power. It shapes the identity and collective sense of life in a place, territory, and space.

Within this framework, *acoustic colonialism* can be characterized as a regime of massive, iterative, and invasive linguistic, ambient, social, and technological mechanisms that alter and disrupt the lifeways of a Native territory. My theorization of acoustic colonialism and its related concepts—the colonial ear, Indigenous/Mapuche interferences, and aural and sonic autonomy—emerges from multiple dialogues and reflections. As soon as I became interested in working on questions of sound and indigeneity, I had to figure out how to situate them in connection to the historical issue of colonialism, since it was a cardinal aspect in my Williche experience as well as in the Mapuche critical discussions I have been involved with since the late 1990s.

Undoubtedly, Mapuche anticolonial critique of the Chilean and Argentine nation-states as oppressive settler powers has been influential in my work. The publication of the book *¡ . . . Escucha, winka . . . ! Cuatro ensayos sobre historia nacional Mapuche y un epílogo sobre el futuro* in 2006 by Mapuche authors Pablo Mariman Quemenado, Sergio Caniuqueo, José Millalen Paillal, and Rodrigo Levil Chicahual marked an important inflection in this process. In 2009, I contributed to the foundation of the Comunidad de Historia Mapuche (CHM), a collective of Mapuche researchers that aimed to rethink historically and theoretically the issue of colonialism as a nodal dimension in the relationship between the Chilean nation-state and the Mapuche People.[6] Engaging in these collective dialogues has enriched my critical formulations on the aural and sonic edges of colonialism.[7]

In a more classical venue, the writings of Frantz Fanon on the role that radio played in the Algeria of the 1940s and 1950s have nurtured my elaborations on voice, sound, and audibility as entangled in political, racial, and cultural clashes between colonizing and colonized forces. In the collection of essays *L'an cinq, de la Révolution Algérienne* (1959), published in English as *A Dying Colonialism*, Fanon discusses how the radiophonic medium was a tool of French "colonial domination," especially in the 1940s. He also explores how radio became an influential site for the connectivity and dissemination of the voices of an anticolonial movement of "liberation" from the first transmission of *Voice of Algeria* in 1956 ([1959] 1967, 69–97). Traversed by my own historical sense of Wallmapu as a territory colonized and occupied by Creole nation-states, these dialogues, readings, and concrete experiences have led me to intertwine my inquiries on sound and indigeneity with theorizations on settler colonialism and critiques of settler-state formations (Speed 2017).[8]

Furthermore, recent works on sound, listening, and acoustics rooted in Indigenous knowledge-based modes of analysis and radical critiques of oppressive colonial, racial, class, and gender systems have also enriched my critique of acoustic colonialism. In this vein, it is worth mentioning the works of some Native scholars from what is today known as "North America," such as Jessica Bissett Perea (Dena'ina), Trevor Reed (Hopi), and Dylan Robinson (Stó:lo), who have made significant contributions to the emergence of a current of Indigenous anticolonial thought in sound studies (Perea 2021; T. G. Reed 2019; Robinson 2020). Indigeneity, race, sound, and power relations have been addressed in multiple ways by several non-Indigenous scholars such as Alejandra Bronfman, Tina Campt, Licia Fiol-Matta, Ana María Ochoa Gautier, Anthony Reed, and Jennifer Lynn Stoever, who are also relevant interlocutors throughout *Acoustic Colonialism* (Bronfman 2016; Campt 2017; Fiol-Matta 2017; Ochoa Gautier 2014; A. Reed 2021; Stoever 2016).[9]

With these points of reference in mind, I position the category of colonialism as a concrete sensorial, experiential, and historical notion. Recentering the discussion of colonialism in these terms allows me to situate my reflection within the context of Wallmapu as an occupied territory, and, accordingly, weave anticolonial or decolonizing aural and sonic politics and poetics of Indigenous autonomy and resistance. Within this plot of power relations, acoustic colonialism manifests in several terrains. Undoubtedly, sonic and aural environments are defined by what we hear or listen to and by the sounds that permeate our everyday lives within a territory, a community, or at home. In this sense, acoustic colonialism occurs first through the deployment of sociolinguistic dominance, marginalization, or exclusion of native languages; or, more radically, through their erasure or elimination. The latter is what some scholars define as "linguicide."[10] Linguistic colonization, undoubtedly, constitutes a critical aspect of acoustic colonial regimes. Second, media systems and related technologies of representation and communication shape the form and content of acoustic ecologies, becoming key agents in the gears and infrastructure of colonial domination. Third, the technological and infrastructural apparatuses of colonizing powers also are key to the deployment of acoustic colonialism. The noise of industrial machinery, airports, hydroelectric plants, forestry machines, and cargo trucks as well as the weaponry of state police have all become invasive agents that occupy the sonic, visual, and perceptual environment of Mapuche life.

The aural and sonic dimensions of language are thus crucial to the fabric of colonial powers. Settler nation-states have marginalized Native languages to impose, officialize, and naturalize the colonizing language as part of their modus operandi. This has historically affected Mapudungun, the Mapuche language that inhabited "acoustic space" (Schafer [1977] 1994, 271) long before the arrival of the noisy linguistic, cultural, social, and political machinery of settler nation-states. Its decline among younger generations of Mapuche has worsened the marginalization of the native language. The Mapuche linguist María Catrileo notes that, even though Mapudungun has survived across time, "according to research conducted by public and private organizations, the number of speakers is decreasing" (Catrileo 2010, 40). Linguist Fernando Zúñiga affirms that "Chilean Spanish dominates private and public spaces; Mapudungun is an unknown language for most of the population. Few speak it, and English takes over more and more spaces, especially virtual spaces, even though not many people understand it and very few people speak it" (2006, 48). Due to its naturalization as the hegemonic language, Spanish, with its inherent marginalization of Mapudungun and other Indigenous Peoples' languages, dominates the country's sonic environment. Linguistic and discursive colonization fractures and permeates many Mapuche ears and speaking bodies, as media outlets, social institutions, and technological forces constantly deploy the sounds of Spanish throughout Native environments. In this complex web of acoustic colonialism, dominance and hegemony take shape through the spread of the settler language, along with the deployment of economic, social, cultural, technological, and media agents over Mapuche territories.

Considering this historical context, I examine the role of sound in Chilean and Mapuche cultural production over the last two centuries, both as a site of racial, patriarchal, and colonial hegemony and as a space of Indigenous interference and agency against the invasive waves of acoustic colonialism. From the mid-nineteenth century to the late twentieth, Chilean literature, radio, and film had a historic role in disseminating distorted visual and sonic representations of the Mapuche. The enduring mediating effects of what I refer to as the *colonial ear*—the entry point for these misrepresentations—reflect the racial, political, cultural, and historic logic of Chile's settler nation-state. In response and resistance to these aural and sonorous figurations of the Mapuche in Chilean settler literature and media, contemporary Mapuche writers, artists, and activists have produced literary texts, radio programs, and audiovisual media of their own. The voices, sounds, and discourses of this Mapuche production constitute alternative sonic registers and testimonials that embody forms of contestation to and enduring the acoustic colonialism that has long dominated the soundscape of *Ngulu Mapu*—or "Lands of the West"—a territory designated in present-day cartography as central and southern "Chile."

NATIVE TERRITORIES: MAPPINGS AND COLONIAL DOMINATION

From the very beginning, auditory and sonic effects have been woven into colonial violence in the southern regions of the continent that was named the "Americas."[11] North of Mapuche territory, in the Andean highlands of Tawantisuyu, the Indigenous ear was first exposed to these effects when the Spaniards arrived in the sixteenth century. In *Nueva corónica y buen gobierno* (1615), the Quechua writer Guaman Poma de Ayala describes the arrival of conquistador Francisco Pizarro and his men in Cajamarca, in the highlands of the Tawantisuyu, on November 16, 1532, like this:

> Hernando Pizarro y Sebastián de Balcázar [Benalcázar]; de como estuvo el dicho Atagualpa Inga en los baños allá fueron estos dos dichos caballeros encima de dos caballos muy furiosos, enjaezados y armados, y llevaban muchos cascabeles y penachos, y los dichos caballeros armados empuntan [de punta en] blanco comenzaron a apretar las piernas, corrieron muy furiamente, que fue deshaciéndose, y llevaban mucho ruido de cascabeles; dicen que aquello le espantó al Inga y a los Indios que estaban en los dichos baños de Cajamarca, y como vido nunca vista con el espanto cayó en tierra el dicho Atagualpa Inga de encima de las andas, como corrió para ellos, y toda su gente quedaron espantados, asombrados, cada uno se echaron a huir porque tan gran animal corría, y encima unos hombres nunca vista; de aquella manera andaban turbados. (Poma de Ayala [1615] 1980, 291–93)

> Atahualpa Inca was at the baths close to Caxamarca when Hernando Pizarro, the general's brother, Hernando de Soto, and Sebastián Benalcázar rode out to visit him on their prancing horses. They were in full armor, magnificently mounted, with plumed helmets and jingling bells. With the pressure of their thighs, they caused their horses to curvet in front of the Indians and soon made a sudden charge toward the imperial party. The sound of bells and the hoof-beats produced utmost consternation. The Inca's bearers fled in terror at the sight of the huge animals and riders careering toward them and the Inca himself fell to the ground from the litter on which he had been sitting. (Poma de Ayala 1978, 108)

Notably, in this account, as Guaman Poma describes it, "the sound of bells [*el ruido de cascabeles*] and the hoof-beats produced utmost consternation" among the Inca and the Natives. In the Quechua language, *mancharikuy* means "to feel scared or afraid of something." The term could well describe the feeling of Atahualpa and the Quechua people at the time. A related Quechua concept that might also describe this feeling is *ayqichiy*, meaning "to frighten off, to make run away" (Laime Ajacopa 2007, 64). The exposure of Natives' ears to the noise (*ruido*) of the Spanish cavalry triggered the experience of *mancharikuy*. This feeling of terror (*espanto*) from the invasive noise reveals the auditory and sensorial trauma that the colonial encounter provoked in the inhabitants of Tawantisuyu. Undoubtedly, the sonic dimensions of the Spanish army's movement had a traumatic effect. Across time, this pattern would be reiterated in many Indigenous settings.

To the south of Tawantisuyu lies another region that appears on colonial maps as the territory inhabited and controlled by the Mapuche, the region that *criollo* (Creole) settlers came to designate, delineate, and represent as "Chile" and "Argentina."[12] Beginning in the late twentieth century, Mapuche organizations and movement leaders have reclaimed this territory as Wallmapu, a land that was free and borderless prior to the establishment of the settler nation-states. In his contribution to *¡. . . Escucha, winka . . . !*, Mapuche historian Pablo Mariman Quemenado claims that prior to settlement in the latter half of the twentieth century, "the territories disputed by the states" of Argentina and Chile had formed part of the vast independent space of the Mapuche people. Mariman Quemenado states, "This [region] was never Chilean or Argentine. We call this the Wallmapu, the Mapuche Country" (2006, 53). Similarly, historian and activist José Millalen Paillal argues that the Mapuche inhabited a vast territory prior to the arrival of the Spaniards: "It is well known that from the early sixteenth century, the groups that make up the Mapuche culture had settled in the vast territorial space that extends from the Limarí River in the north to the large island of Chilwe (Chiloé) in the south, from the west of the Andes and the eastern slopes of the cordillera, to the northern and central areas of Newken province, and to the south of the current providence of Mendoza, Argentina" (2006, 19).

The Mapuche considered the "western slope" toward the Pacific Ocean, that is, the regions eventually claimed by the Chilean nation-state, to form part of Ngulu Mapu (Lands of the West). They referred to the lands from the eastern slopes of the Andes to the Atlantic Ocean, the region later claimed by the Argentine nation-state, as Puel Mapu or "Lands of the East" (Millalen Paillal 2006, 36–37). Sixteenth-century Spanish colonial agents endeavored "to discover, conquer and populate" these Mapuche territories (Valdivia [1545–52] 1970, 52). Between 1535 and 1537, the conquistador Diego de Almagro began this project in earnest; however, it was only with Pedro de Valdivia's more aggressive invasion in 1540 that the Spaniards attempted to incorporate the Mapuche territory into their newly established Viceroyalty of Chile. When Creole elites founded the "Republic of Chile" in 1810, the Mapuche people still widely controlled Ngulu Mapu, particularly between the Bío-Bío and Tolten Rivers. This posed a challenge for the geopolitical rule of the newly formed Chilean settler state, which then plotted its own colonial agenda.

In line with the history of Mapuche resistance and the stance of the Mapuche movement of the late 1980s, the authors of *¡ . . . Escucha, winka . . . !* understand the term Wallmapu as a political and historical vindication of the entire Mapuche territory. In Mapudungun, Wallmapu is related to the concept of the universe as a whole. By the late twentieth century, however, Mapuche leaders and organizations began using it to refer to País Mapuche (Mapuche Country), or Nación Mapuche, a semantic shift that has become part of the battle to reclaim the names and representations of territories that had been subsumed by the denominations "Chile" and "Argentina." In *¡ . . . Escucha, winka . . . !* Mariman Quemenado inserts two hand-drawn maps (see figures I.1 and I.2) to graphically and figuratively represent Wallmapu (2006, 60, 77).

Mariman Quemenado's first hand-drawn map covers a whole page.[13] Visually, Puel Mapu (Lands of the East) is placed above, while Gvlu Mapu (Lands of the West) appears below. This first map, titled "Wall Mapuche / Nación Mapuche," features the traces of the Fvxa Mawida (the Great Cordillera). The northern side includes a series of *pewen* (native mountain trees); names of places; visual traces and names that mark the veins of rivers; figures of men on horses or simply standing; a large *ruka* (Native house); inscriptions of several *ñimin* (symbolic Mapuche designs). In differently sized letters, Mariman Quemenado draws the names of the different Wichan Mapu (lands-peoples) who form part of the Wallmapu. The Wichan Mapu are written in capital letters: Nagche, Lafkenche, Wenteche, Pewenche. So are neighboring peoples such as the Tewelche to the south and the Rankvlche and Cadiche to the north in Puel Mapu. In smaller letters, next to local territorial identities, other Wichan Mapu are named: Williche, on the southern ends of Gvlu Mapu

FIGURE I.1 · Map of "Wall Mapuche Nación Mapuche." Hand-drawn by Pablo Mariman Quemenado, early 2000s.

and Puel Mapu; and the Pikunche on the northeast, toward Puel Mapu. Then we turn a few pages to get to the second map.

The second map occupies less than half of the page, with the east visually placed on top and the west below. Under the title "Wall Mapu País Mapuche," it visually depicts the nonhuman lives of Wallmapu, naming, tracing, and drawing mountains, rivers, lakes, different regions, urban settlements and places, some pewen trees on the cordillera, and, notably, several nonhuman animal figures. At the bottom, is the Fuxa Lafken (Pacific Ocean). If we look back at the main text, we read and see a subtitle that marks the broader sense of these cartographic inscriptions: "The Mapuche Political and Territorial Independence" (Mariman Quemenado 2006, 77).

¡ . . . Escucha, winka . . . ! was the first book on the territorial history of "South America" to include such a representation. Previously, all Argentine and Chilean cartography superimposed the geopolitical borders of the Creole nation-states onto Native territory. Provoked by discussions on territoriality at the Instituto de Estudios Indígenas at Universidad de La Frontera in Temuco between 1999 and 2002, Mariman Quemenado drew the maps to visualize what, from a Mapuche perspective, had to look different from the geopolitical figurations of

FIGURE I.2 · Map of "Wallmapu País Mapuche," including Wallmapu's ecosystem. Hand-drawn by Pablo Mariman Quemenado, early 2000s.

"Chile" and "Argentina." Recalling discussions among Mapuche activists, Mariman Quemenado states, "Toward the end of the 1980s, we started to speak of the *País Mapuche.*" He challenged himself to draw maps that reflected an emancipatory politics of Wallmapu. Mariman Quemenado's drawings revert to an older version of mapmaking to depict the diverse human and nonhuman land relations that constituted Mapuche life. Borrowing images from different sources and placing them within a Mapuche cartographic narrative, he aimed to represent a territory that "is a world in which all those diversities that are there with us coexist" (Mariman Quemenado, interview by author, 2023).

Mariman Quemenado thus returns to an ancient figurative cartographic representation to challenge the more abstract geometry of the "modern" mapmaking of colonizing, settler nation-states. In representing the heterogeneity of Wallmapu, Mariman Quemenado's maps capture an ontological, cultural, political, and territorial conception of the greater "País Mapuche." According to Mariman Quemenado, Wallmapu constituted "an alliance of Wichan Mapu more than a state or a single community" (interview by author, 2023). In my reading, his cartographic drawings also embody a desire to inscribe an important ontological, ecological, and cultural Mapuche principle, namely, Itro Fill Mongen, which refers to the diverse tangible and intangible relationships and multiple lifeforms that inhabit and underlie the ecosystem of Wallmapu. Considering these dimensions, Mariman Quemenado's maps embody a politics and aesthetics of Indigenous Mapping, thus interfering with geo-colonial logics of misrepresentation on ancestral lands.

The contemporary Mapuche movement has embraced the mapping of an "independent" Wallmapu to unsettle the cartography of the Creole nation-states and reestablish the memory of the self-governing territory that occupied the same space as the settler nation-states. To the south, Wallmapu bordered the lands and waters of what today is called "Patagonia." To the north, it bordered the domains of Tawantinsuyu, which generated geopolitical tensions due to the expansive designs of the Inca Empire. In their efforts to delineate the understudied and porous northern frontier of Wallmapu, Millalen Paillal and Mariman Quemenado made use of archival records and Mapuche oral history. While Millalen Paillal locates the Wallmapu-Tawantisuyu border at the Limarí River, in the contemporary Chilean province of Coquimbo, Mariman Quemenado offers a slightly different rendering of the northern limits of Wallmapu. He suggests that when the Europeans arrived in 1536, the borders of the Mapuche Nation extended from the island of Chiloé to the Copiapó River, which runs through the province known today as Atacama, north of Coquimbo.

These historical accounts concur that the Pikunche (Mapuche of the North) lived in the region between the Copiapó and Limarí Rivers, where they coexisted with the Diaguita people. What is less clear, however, is whether the Mapuche were in full control of the northern territories of Ngulu Mapu up to the Limarí and Copiapó rivers. Native chronicles of Tawantinsuyu claim that by the end of the fifteenth century, Inca expansionism had undermined the territorial sovereignty of the Pikunche and Diaguita, as the Inca Yupanqui's forces penetrated as far as the Maule River region, south of today's Santiago. In his *Comentarios reales* (1609), the Quechua-Spanish chronicler known as "the Inca," Garcilaso de la Vega, recounts military campaigns into the valley called *Chili.*[14] This account included Inca-era expansions into Ngulu Mapu, as well as incursions later carried out by the invading Spaniards (Vega 1966, 445–62). After reducing the province of Copayapú (Copiapó) to his rule, King Inca Yupanqui "sent forward another army of ten thousand men, with all necessary supplies, who were ordered to assist the first two armies," undertaking a new expansion that enable them to press "on a further eighty leagues" and annex "the valley of Coquimbu" to their empire (Vega 1966, 447). The Inca forces then moved south to the province they called "Purumauca," where they met Mapuche resistance. To avoid a prolonged war, the Inca Yupanqui ordered his men to halt their expansionist endeavors. Inca Garcilaso writes, "With these instructions the Incas made an end to their conquests in Chile, strengthened their frontiers, set up boundary marks and fixed the Maule River as the furthest limit of their empire toward the south. They attended to the administration of justice, the royal estates, and the property devoted to the Sun, the special benefits of their vassals. The latter accepted the rule of the Incas, and their privileges, laws, and customs with true affection, and lived under these laws until the Spaniards entered Chile" (Vega 1966, 450).

By the sixteenth century, the once-sovereign territory of the Pikunche that Garcilaso referred to as "the kingdom of Chile," was subjected to at least partial Inca control. When the Spaniards arrived, it was the Maule River, not the Copiapó River or the Limarí, that marked the northern border of Wallmapu. Although Inca expansionism had altered northern Mapuche territory and weakened the Pikunche community system, the colonizing invasion of the Spaniards met serious resistance. The Spanish army led by Pedro de Valdivia consisted of two hundred men, including Indigenous warriors and servants recruited in Tawantinsuyu. In a letter dated on September 4, 1545, and written in the northern city of La Serena, Valdivia recounts "not having seen between Peru and here any Indian in peace." He met with "rebels of the *mantenimientos* everywhere" (Valdivia [1545–52] 1970, 37). This resistance was a preamble to what Spanish conquerors encountered in the heart of the Mapuche territory south of the Maule River. There, the Arauco War lasted from the mid-sixteenth century through the seventeenth. To undermine the sovereignty of Ngulu Mapu, the Spaniards established forts and small urban settlements between the Maule River and Chiloé. The Mapuche resisted the expansion of Spanish colonization until the eighteenth century, when after a series of peace summits (*parlamentos*) with Spanish officials, they were finally able to negotiate significant control over their territories.[15]

The formation of the independent republics of "Chile" and "Argentina" led by the Creole elite in the early nineteenth century, however, led to severe regression of Indigenous territorial sovereignty. The "Campaign of Pacification of the Araucanía" in Ngulu Mapu (from 1862 to 1883) and the "Campaign of the Desert" in Puel Mapu (from 1878 to 1885) were the most intensive state-sponsored efforts to colonize Mapuche lands through war (Mariman Quemenado 2006, 101–13). In some areas, Christian missionization became an expedient added to this process. In Chile, the paradoxically called "Campaign of Pacification" started in 1862 through the Chilean state-sponsored military invasion of Ngulu Mapu, the Mapuche territory. Its launching was led by Colonel Cornelio Saavedra through the invasion of the plains of Angol (Province of Malleco), thus paving the way for the settlement of economic, administrative, and ecclesiastical agents on Mapuche lands; a process of land dispossession that—as a structural or systemic pattern of domination—started by then but continued throughout the past century, and keeps up to this day. At the core of the Chilean colonization enterprise has been what Australian anthropologist Patrick Wolfe has defined as a key component of "settler colonialism," namely, "access to territory" and subsequent dispossession and occupation of native lands by settlers who, in the end, had "come to stay" (2006, 387–88). During the Pacification Campaign period, settler colonizers—a good number of Chilean Creoles and others coming mostly from Germany—arrived in Ngulu Mapu with a "justification" similar to that of their Anglo settler peers in Aboriginal Australia, namely, "we could use the land better than they [the Natives] could" (Wolfe 2006, 389). Several studies of the period confirm that settler land occupation constituted a cardinal aspect in Wallmapu, as part of the expansive geopolitics of both the Argentine and Chilean states. For the case of Ngulu Mapu, since the Pacification Campaign onward, the aggressive Chilean state occupation of the native territory ranged from forced territorial removal by direct military action to fabrication of the consent of communities under the pressure of a threating militarized environment or through fraudulent paperwork in notarial offices or the subterfuges of the gelatinous Chilean law (Correa Cabrera, 2021).

Added to this, a recent study by Mapuche historian Pablo Mariman Quemenado highlights another strategic dimension of colonialism peculiar to the nineteenth-century Mapuche experience. He demonstrates that, besides the question of the "native land," the livestock wealth of many Mapuche communities triggered colonial greed over the territories of Ngulu Mapu and Puel Mapu. Before, and even during, the period of the Chilean and Argentine colonial invasions, ownership of a significant amount of livestock—*kulliñ*, in Mapudungun—formed the base of Mapuche people's wealth.[16] Chilean settler colonization indeed entailed not only the settler occupation and reduction of our territories but, as an important historical trait of colonialism in the Mapuche context, the appropriation or theft of cattle and other goods from economically prosperous communities at the time (Mariman Quemenado 2023). Evidently, theft of livestock from the Mapuche was an attractive path of fast capital accumulation for the colonizers, and, from the colonized perspective, the beginning of the sad history of economic weakening and progressive impoverishment of Mapuche communities. Thus, both key correlated factors, colonial possession of Indigenous lands and appropriation of Mapuche livestock wealth were certainly at the core of the tangled power relations incubated under Chilean colonialism—an invasive colonial deployment that, besides geopolitical, economic, racial, and cultural domination, would also imply a fundamental alteration of the acoustic ecology of Ngulu Mapu.

ACOUSTIC COLONIALISM AS TRAUMA AND BATTLEGROUND

Historical attempts to occupy Mapuche territory have entailed the disruption of its sonic environment. The colonial invasion of Tawantinsuyu put ears and bodies into a state of fear: The noise of the Spanish army "scared the Inca and the Indians" (espantó al Inga y a los Indios) (Poma de Ayala [1615] 1980, 108). Then the colonial machinery of sound and terror that Ngulu Mapu had endured during the centuries of Spanish colonialism became the genocidal "Campaign of Pacification" of the nineteenth century. The Chilean military invasion entailed a collective shock, as illustrated in several passages of *Crónica de la Araucania* by Creole army official Horacio Lara, a protagonist in the Chilean colonizing enterprise during the 1880s.[17] Lara writes,

> El 2 de diciembre llegaba el coronel Saavedra con su division a Angol, i desde su campamento tomaba todas las medidas que estimaba prudentes para calmar la susceptibilidad de los indios alarmados con la presencia de tropas en su territorio. Era verdaderamente penoso presenciar los llantos i esclamaciones de dolor de las mujeres araucanas al ver que se instalaban nuestros soldados en sus posesiones de donde huian despavoridas a los bosques con sus hijos. (Lara 1889, 265)

> Colonel Saavedra arrived in Angol with his division on December 2 and took the measures he judged prudent to calm the Indians, who were alarmed by the presence of troops in their territory. It was truly painful to witness the Araucana women's weeping and exclamations of sorrow as our soldiers seized their lands from which they fled in terror to the forests with their children.

This passage evokes the emotional, physical, and acoustic trauma that Native people faced during the Spaniards' colonial invasion. Three centuries after the 1532 events in Cajamarca, and in the context of the Chilean nation-state capitalist expansion, Lara's narrative recounts a similar deployment of colonial violence in Ngulu Mapu, with "alarmed" Indians and Mapuche women who "fled in terror." Even today we witness this type of "state terrorism" against Mapuche resistance in Ngulu Mapu, a deployment of colonial state violence that began with the Chilean military invasion of our territories in the mid- and late nineteenth century.[18] Lara's reference to the Mapuche women's "weeping and exclamations of sorrow" highlights a traumatic sonic environment, with colonized bodies once again frightened and in tears.

In Lara's chronicle, an acoustic settler logic imbues the terrain of language. His reference to Mapuche women as "Araucanas" reiterates a rhetoric of denomination and representation. Historically, the Mapuche became known as "Araucanians" through the filter of the colonial ear. The Inca Garcilaso uses the term *Araucos* to identify the inhabitants of a region south of the Maule River known to the mestizo Inca writer as "the Province of Araucu" (Vega 1966, 314–15). The Inca's use of this name seems to have coincided with the Spaniards' use of the term. By the mid-sixteenth century, foreigners commonly used the term *Araucos* for the Natives of the lands that contemporary Chilean maps designate as the provinces of Bío-Bío, Arauco, Malleco, and Cautín. Writers in the service of the Spanish Crown consolidated this exogenous labeling.

An antecedent of the use of this lexicon is *La Araucana*, the epic poem written in the 1560s by Spanish scribe Alonso de Ercilla y Zúñiga (Madrid, 1533–94). *La Araucana* eulogizes the "Araucanians" as warriors of great "physical" attributes. The settler Chilean state reiterated this Araucanist rhetoric of colonial literature in its written and oral discourse from the early nineteenth century. *La Araucana* became a canonical text in the school system and a frequent reference for Chilean scholars, writers, educators, and public leaders over the past two hundred years. Generations have been forced to memorize and recite verses from *La Araucana*, which has seeded our auditory imagination with the term *Araucano* to misname the Mapuche. Words such as *araucanos, araucanas, Arauco*, and *Araucanía* have had long lives in the linguistic ethnogeography of representation and naming in Chile. However, in the late twentieth century, leading Ngulu Mapu Indigenous organizations began to challenge this wording, preferring the self-identificatory term *Mapuche*.

What exactly is the origin of the colonial nomenclature of Arauco? Mapuche linguist María Catrileo has delved into the phonetic and morphological genealogy of *araucanos* and the adjacent concept of *Arauco*. She finds the likely source of the term *Arauco* in the Mapudungun words *rag* (clay) and *ko* (water), the combination of which could be translated as "water among clay" (2010, 23). The clay that colors the tributaries and streams of the Maule River in northern Ngulu Mapu no doubt explains the use of this composite term. Catrileo suggests that through phonetic association in the ears of foreigners the regional use of *Ragko* became *Araucu* among Quechua speakers and *Arauco* to Spanish colonizers. The conversion and assimilation of Indigenous names and terms into phonemes more legible to the ears and vocal repertoire of Spanish speakers was (and remains) a dominant pattern. Along with the technology of writing, the spread of denominations filtered through and formatted by the phonetics of the colonizers' language increasingly populated the acoustic ecology of Ngulu Mapu.

Since the era of Spanish colonialism, the Mapuche have generated their own filters for naming and representing the settlers. In an October 15, 1550, letter to Charles V recounting incursions into the "Arauco" region, Pedro de Valdivia cited an example from the colonial settlement of Concepción. The Natives, he wrote, "call us *ingas*, and our horses *hueques ingas*, which means 'sheep of the ingas'" (Valdivia [1545–52] 1970, 157). Over time, the Mapuche identified all foreigners as *wingka* (similar to the phoneme *inga*). According to Mapuche scholar María Catrileo, *wingka* is a generic term referring to the "non-Mapuche person or people" and sometimes can be used in a neutral or even friendly tone for non-Mapuche (2017, 45). The independent Mapuche researcher Martín Alonqueo, however, situates the term within a history of "hatred, rancor, and contempt that arose in the epoch of the discovery [and continues] into the present." Alonqueo claims that *wingka*, "etymologically, according to the Mapuche language, means thief, highwayman, invader. It derives from the verb *winkün*, which means to steal and invade by force, entering without authorization" (Alonqueo 1985, 143). Alonqueo offers no evidence of this "etymological" framework, but he does present a sociolinguistic gloss for the Mapuche's historical association of the settler with a thief, invader, or liar.

The semantic layers of wingka performatively resonate in a variety of contemporary speech settings. The title of *¡ . . . Escucha, winka . . . !* illustrates how the verbal and phonetic field constitutes a space of struggle for what is written, spoken, and heard: a centuries-long battle over language and enunciation. The Native subject—labeled as "araucano" or "indio"—has responded with multiple counter-denominations, including "wingka"! Likewise, that response compels the settlers to open their ears and engage in listening.

COLONIAL CONTINUITY, NEOLIBERAL SCENE

In writing this book, I had in mind the history of wingka settlements that have affected the Mapuche territory and people from the first Spanish invasion to the establishment of the Chilean and Argentine settler colonial states. This history continues to resonate as an invasive process that alters the acoustic ecology of Wallmapu and suffuses it with non-Indigenous sonic forces, systems, and lifeways. An intricate history of "colonial entanglement" (Dennison 2012) often mediates ways of living, experiencing, and exercising indigeneity in the terrain of sound. As an "acoustic territory" (LaBelle 2010), Wallmapu has been subjected to the dynamism and intricacy of sonic flows. Sound is often porous, not always allowing for well-defined or fixed acoustic boundaries between what is colonizing and what is not. Yet, within the complexities of this entangled soundscape, the Mapuche have managed to forge forms of persistence, self-determination, self-affirmation, and agency.

As part of the ongoing struggles for a "Free Wallmapu," since the late 1980s many Mapuche leaders and communities actively embraced the principle of autonomy to shape a politics of liberation from the rule of the Chilean nation-state and settler capitalist agents. Mapuche-Lafkenche scholar Héctor Nahuelpan Moreno links autonomy to a Mapudungun-based notion, *kisugunewün*, which has a much longer history in Wallmapu. In his words, this Mapuche concept is "the ability of an individual or collective self 'to be in control of itself' or 'to govern themselves'" (Nahuelpan Moreno 2013, 10). In the context of Abiayala, Mapuche people have not been alone in embracing this Indigenous horizon of autonomy. For example, since its emergence in Chiapas, southern Mexico in the mid-1990s, the Indigenous Zapatista movement has promoted a politics of autonomy understood as a way "to do things ourselves, with our own resources and our own ideas" (Mora 2017, 6). Such movements have encouraged a transnational discussion on the possibility of restituting Indigenous territorial and communal lifeways outside the jurisdiction and dominion of settler states. In the terrain of sound, a politics of autonomy can also become a critically generative concept.

In what today is called "North America," or Turtle Island—to use one of its main ancestral denominations—"sovereignty" has been a prevailing category to articulate politics of Native self-determination. Certainly, this notion has troublesome ties to a Western patriarchal sense of central power and authority, be it an emperor, king, or state figure. Trevor Reed (Hopi) problematizes this historical genealogy of the concept, as it "originally referred to the divine authority of European kings" and then, by the late eighteenth and early nineteenth centuries, it became the recolonizing entity of the "newly formed nation-states" (2019, 511). Based on a critical examination of the concept at play and explicitly questioning what he characterizes as "the human-centered, totalizing sovereignty model of the American settler-state" (2019, 525), Reed redefines questions of authority, peoplehood, and territory to envision a more situated and relational Indigenous sense of sovereignty—or, more specifically, "sonic sovereignty." He points out that "in Indigenous communities, sovereignty exists within ongoing generative relations, rather than being prescribed by a single, monolithic authority or text" (T. G. Reed 2019, 513). Furthermore, "When humans are de-centered and where sovereignty is not totalized around any one entity, as may be the case for many Indigenous sovereigns, it seems that the role of sonic authority may be all the more important" (T. G. Reed 2019, 525). This is the case, for example, with Indigenous songs meant to resonate in territorial and communal relations, wherein "the creation and performance of Indigenous song should be understood as an act of sovereignty" (T. G. Reed 2019, 526).

This recontextualization of sovereignty, especially in the realm of cultural life, resonates well with the political horizon of *kisugunewün*, or autonomy, as discussed in Wallmapu. Yet Mapuche struggles for autonomy, particularly through acts of land recovery, have been detached from the shadow of statehood. They have refused colonial mediation, or what political scientist Glen Coulthard (Yellowknives Dene) critiques as the politics of "colonial recognition" (2014, 25–49). Thus, my approach is driven by a Mapuche sense of *kisugunewün* that constitutes an ongoing collective refusal of settler-state subjection or condescension. It is a standpoint critical to the politics of land restoration embraced by many communities in Wallmapu. Definitely less haunted by the specter of "sovereign" authority, autonomy works for a local, territory-based polycephalic society like the Mapuche. The Mapuche never endorsed the idea of a single, central authority, as was the case with other Indigenous societies that became empires, for example. Therefore, present-day attempts to impose state-centric "Indigenous" policies—be they "multicultural," "intercultural," or "plurinational"—on Mapuche collective life do not and will not work for us. In the realm of acoustic ecologies, it is strategic to envision a Mapuche politics and poetics of aural autonomy.

Aural autonomy certainly involves a sense of Indigenous emancipation in the terrain of the sonic, or "sound relations," as conceptualized by Native Alaskan scholar Jessica Bissett Perea (2021). But it also engages the broadest sense of "aurality" (Ochoa Gautier 2014) by addressing liberatory practices of listening and corporeal and territorial reverberations that form the ancestral and present-day sonic environments of Wallmapu, including the multisensorial and polymorphous subjective and intersubjective experiences that constitute it. Aural autonomy is an ongoing dream. It is a Mapuche dream. It is therefore part of the dream of a Free Wallmapu that takes the historical shape of a political and cultural force that invigorates, challenges, and puts in tension the liberatory possibilities as well as the limits of artistic and cultural practices.

With this horizon of aural agency and liberation, and challenging our own entangled colonial histories, I analyze how the Mapuche have used literature, arts, and the media to formulate responses to acoustic colonialism. To contextualize this story, the opening chapters offer an overview of Chilean literary and settler cultural representations of the Mapuche sonic environment. They explore the role of the colonial ear and media in staging acoustic colonialism and its derivative politics of misrepresentation up through the twentieth century. Subsequent chapters analyze Mapuche literary and media interventions that developed in the context of the neoliberal and extractive capitalism of the late twentieth and early twenty-first centuries.

The invasion and occupation of territory to the detriment of Indigenous people are fundamental to colonialism. This reality intensified further under the economic, political, and cultural regime of neoliberalism that dominated Chile after the "structural adjustment" of the mid-1970s.[19] The collusion of the military dictatorship led by General Augusto Pinochet and monetarist policymakers dismantled the state-based economy and paved the way for a market-oriented regime that privileged private capital. The effects on Ngulu Mapu were dramatic. According to historian Thomas Klubock, in the 1930s, "state institutions began to invest heavily in pine plantations and to funnel assistance to landowners who cultivated tree plantations" as part of mixed state-private "development" in southern Chile (2014, 122). By then, for governmental agents, Monterey pine was an "alternative to native forests, whose extraordinary heterogeneity render them difficult to log profitably" and this pine species "could be grown cheaply and quickly to supply expanding national and international markets for lumber and paper pulp" (Klubock 2014, 20). Then, in 1974, Pinochet's military regime enacted Law 701 to promote "the forestation of rural surfaces with exotic monocultures of pine and eucalyptus trees" by subsidizing up to 75 percent of private entrepreneurial investments through tax exemptions and other provisions (Levil Chicahual 2006, 232). In the emerging neoliberal scenario of the late 1970s and subsequent decades, along with eucalyptus, it was the turn for another pine species, *Pinus radiata*. Law 701 marked a strategic turn in this economic path by providing it with the allegedly "legal framework" to solidify aggressive and invasive expansion of the forestry industry in Mapuche territory. Across time, state and corporate investments in artificial forestry have been part of the strengthening of this industry in the lands of Ngulu Mapu, thus expanding the extractive economic pattern that has shaped Chilean capitalism since the nineteenth century.

By the early 2000s, corporate monoculture exceeded two million hectares in the geographical zone that encompassed the areas known in contemporary Chilean geopolitics as the regions of Maule, Arauco, Malleco, Cautín, Los Ríos, and Los Lagos, that is, most of Ngulu Mapu, or what is today "southern Chile." The expansion of artificial forestry benefited the "economic groups" that rose in the wake of Pinochet's takeover: capitalist conglomerates driven by short-term profits to supply lumber to foreign markets at the expense of unbridled environmental, social, economic, and cultural degradation in Ngulu Mapu. Traveling southward from the Maule Region (or northward from Los Lagos Region), the landscape is visually enveloped by artificial forestry along highways, freeways, and local roads: a homogeneity of tones and shapes that oppresses the very *mapu*—the land, the environment as a whole.

Mapuche scholar and artist Francisco Huichaqueo captures this ecological shock in the video *Mencer Ñi Pewma* (Mencer's Dream, 2011). As the voice-over recites,

> Robles huachos, laurel solitario, pewen extinguido,
> pinos, pinos, pinos, pinos, pinos, pinos.
> ¡Cómo llora el viento anacleto! / ¡Cómo cuentan sus millones!
> Pinos, pinos, pinos, pinos, eucalyptus, eucalyptus, eucalyptus.
>
> Orphan oaks, solitary laurel, pewen extinguished,
> pines, pines, pines, pines, pines, pines.
> How the wind cries anacleto! / How they count their millions!
> Pines, pines, eucalyptus, eucalyptus.[20]

As a companion to the visualization of pine plantations, the elegiac tone of these verses highlights the agony, solitude, and extinction of the native forest's ancestral trees (oak, laurel, and pewen). The first verse constitutes a necrological inscription, an obituary of environmental destruction. It denounces what Mapuche-Tehuelche leader, writer, and intellectual Moira Millán calls *terricidio* (terracide), the "form of destroying life in all its ways" constituting a "continuous aggression against the cosmic order" (2024, 163–64). In Wallmapu, terracide has been historically deployed as the systemic attack against Earth (Mapu) by predatory colonial, capitalist, masculinist, and militaristic state forces and private agents. This ecologically devastating process is made evident by the image of native trees—"Orphan oaks, solitary laurel, pewen extinguished"—that seem to agonize in the face of "the vastness of monocultural forest plantation" (Gómez-Barris 2017, 72). Forest monoculture takes the form of an iteration of "pines" and "eucalyptus," engendering a landscape characterized by a "sameness of tones." The green of radiata pine and eucalyptus creates a monochromatic atmosphere that, in the gaze of Huichaqueo's speaker, turns nightmarish and gray, cinematically black-and-white. The exclamation marks in the third verse contrast the tears of the wind with the profits of the forestry industry. While the "wind" figure is a synecdoche for the suffering natural environment, "anacleto" refers to the Italian Chilean entrepreneur Anacleto Angelini, whose logging company (Celulosa Arauco y Constitución, known as CELCO or ARAUCO) dominates forestry, wood pulp, and engineered lumber industries in the region. Not only is Anacleto Angelini addressed as if he were present, but also, through the sign of interjection, he is called on to listen to and be accountable for the weeping of "the wind." By using the poetic form of an apostrophe with its exclamation, these verses are directly aimed at "anacleto" as the persona who most symbolizes corporate enrichment at the expense of local communities and the environment in Chile.[21]

Besides benefiting an entrepreneurial elite and exacerbating socioeconomic disparities and precarity in the region (Hofflinger et al. 2021), artificial forestation in the age of neoliberal and extractive capitalism has created the conditions for environmental catastrophe and the destruction of Indigenous lifeways. Studies reveal that the desiccation of soil, deterioration of aquifers, and seasonal water crises now common in Mapuche territory are directly linked to pine radiata and eucalyptus monoculture. Summer droughts are recurrent now in many communities of the Malleco and Araucanía regions. Corporate terracide in Ngulu Mapu has attacked the *mawida* (native forest) and *ko* (water) that are critical to Mapuche subsistence economies, spiritual life, ceremonial practices, and health system. For us, many spiritual protectors and ancestors dwell in the native forest. The native forest's streams nurture the growth of medicinal plants. And its springs (*menoko*) play a central role in the labor and lives of the Machi, the main Mapuche spiritual and medicinal authorities in communities.

The ecology of Mapuche life is in shock. Huichaqueo's video poem makes us hear the weeping of one of its vital forces: "¡Cómo llora el viento . . . !" (How the wind cries . . . !). In contemporary Ngulu Mapu, the wind explodes in tears as the melodic murmurs of streams become less and less audible and the subtle sound of burbling springs ceases to exist. With the disappearance of ancestral forests, native birdsong has become harder to hear as well. In addition to the impact of the necroeconomics of the forestry industry and its monoculture, as Huichaqueo denounces, Ngulu Mapu has suffered from several hydroelectric dams, built since the 1990s in response to the voracious energy demands of contemporary urban capitalism. National and transnational private capital dominate Chile's energy-extraction business, which is highly concentrated in mega-corporations such as ENDESA, Colbun, and Aes Gener, corporations that "besides controlling the production of electricity, own

90 percent of water rights" (Colectivo Editorial Mapuexpress 2016, 51). Sponsored by Chilean state policies and agencies, the spread of hydroelectric dams across Ngulu Mapu has led to the corporate seizure of huge tracts of land and led to the forcible removal of large numbers of people and large amounts of water and nonhuman habitats.

Like many other such settler projects on Indigenous lands, the installation of the hydroelectric dams in Ngulu Mapu has been an exercise of colonial violence against what the Mapuche call Itro Fill Mongen. This term constitutes a key Mapuche life concept and principle to describe and define the Mapu as an ecosystem of diverse human and nonhuman lives, including its soundscapes. The deployment of hydroelectric dams, the forestry industry, and similar corporate capitalist invasions of Mapuche lands have imposed and fostered monotony over polyphony, paving the way for what I would call an *acousticide*.

ALLKÜTUN

In Mapudungun, *allkütun* means "to listen attentively, to pay attention to," according to Mapuche pedagogue Clorinda Antinao Varas (2014, 105). The concept invites an engaged mode of listening. It suggests an attentiveness that places one in a conversation (*nütram*). This relational mode of listening nurtures my approach to literature, media, and their contexts. While the making of sounds and the act of listening are figuratively staged in a literary text, they become much more literal in radio, audiovisual works, performances, and songs. The methodological principle of allkütun enables me to actively dialogue with these varied forms of expression. I rely on this aural principle to guide and shape my interpretative analysis, and to help me uncover and interweave the natural, technological, and urban environments that reverberate through the wide range of sonic media I read, watch, and listen to.

Considering the multiplicity of texts, genres, and media discussed in this book, the concept of allkütun also helps tune my ears to engage in conversation with a wide range of practices that run from the figurative "listening" of literary texts to the more direct and literal sonic language of a radio show. "Listening attentively" to texts leads not to a literal auditory practice but, rather, to a figurative sense of listening mediated by the graphic and visual omnipresence of writing. Reading takes primarily the shape of an intensively visual act, a concentrated mode of seeing.

However, literary analysis also requires interpreting traces of sound in written texts. In her discussion of listening, aural registers, and voice, music scholar Ana María Ochoa Gautier has highlighted the role that the "spectrality of sound" plays across different modalities of writing and inscription (2014, 7–8). Whether they appear spectrally or more literally through the phonetics of a text, auditory and acoustic dimensions constantly intermingle and reverberate in language and representation. These dimensions of literary language led semiotics-based scholars of the 1970s to coin the term *phonotext*, a notion still strongly attached to textualism.[22] Contemporary literary critics have elaborated important approaches to questions of voice and sound, particularly in written poetry (Masiello 2013; Perloff and Dworkin 2009; Stewart 2002). Other scholars have theorized the public delivery and recitation of poetry, as embodied in the American critic Charles Bernstein's call for "close listening" (1998). Challenging the boundaries of writing and stressing considerations of listening in literary studies, these contemporary approaches attempt to delve into the sound sphere of poetry and language.

Listening attentively also entails engaging in conversation, collaboration, cross-fertilization, and the principle of reciprocity. Research for this book involved extensive fieldwork, including numerous trips to Ngulu Mapu to conduct archival investigation, dialogues, and oral interviews with writers, audiovisual artists, scholars, radio broadcasters, and other cultural and media activists in the region. Although traditionally associated with the social sciences, the idea of "fieldwork" has become more relevant in the humanities in recent years, amplifying the possibilities of engagement in local and community settings.[23] In my case, as a Mapuche-Williche scholar born and raised in Ngulu Mapu, fieldwork implied a dimension of reconnection; each "research trip" involved the sense of return, renewal, and recommitment to Mapuche cultural and political life. In this process, the practice of allkütun has been essential. It has encouraged me to sharpen my ears and get involved in multiple forms of dialogue with people and with the environment. Within an Indigenous sociocultural framework, "attentive listening" relates to another central notion for the Mapuche: nütram, or conversation. My research, then, takes root in these conversations. The methodological practices of allkütun and nütram entailed open interview formats but also myriad recorded and unrecorded conversations, planned and unplanned engagements, and collaborations over almost two decades.

At the same time, the practice of "attentive listening" can also lead to troublesome processes that end up serving colonial and oppressive practices and interests. Who exercises allkütun? To what end? Discussing the nexus between listening and espionage, French essayist Peter Szendy argues that the deployment of "auditory surveillance" is a key "matter for spies" (2007, 24–25). "Attentive listening" can benefit colonizing agendas when governmental agents, states, empires, or otherwise powerful individuals or interest groups undertake espionage on Indigenous territories to exercise surveillance and counterinsurgency.

A similar critique informs the concept of "hungry listening," coined by music scholar Dylan Robinson. Robinson came up with this notion in relation to the history of settler arrival in what is today called Canada and how the settlers' "states of starvation" for food transmuted into "hunger" for gold and, ultimately, for Indigenous lands, knowledges, and cultures. He thus juxtaposes the logics of appropriation, extraction, consumption, assimilationism, and standardization. For Robinson, hungry listening "privileges a recognition of palatable narratives of difference," while at same time disciplining the ear within Western "standardized features and types." What interests me is the extractive dimension of Robinson's "hungry listening," the settlers' borderless eagerness "to learn and dig into" Indigenous territories, both materially and symbolically (2020, 48–51). In the Mapuche historical experience, there is a long tradition of avid listeners driven by extraction, domination, and colonization. This tradition includes Christian missionaries who sharpened their ears by learning Native languages, legions of Western scholars who immerse themselves in Indigenous conversations to extract knowledge on behalf of preestablished agendas, and state intelligence agents who similarly sharpen their ears to obtain strategic information from Native linguistic and social codes. As Mapuche scholar Héctor Nahuelpan points out, by the second half of the nineteenth century, an army of Capuchin clergymen arrived in Ngulu Mapu to immerse themselves linguistically and culturally in the Native environment, much as the Jesuits and other Franciscans had done in previous centuries to advance their enterprise of religious and spiritual colonization. Evangelization operated as a sort of de facto cultural espionage. Indeed, for Capuchin missionaries a key step in materializing a Christianization process that could touch "the 'heart' of Mapuche families and children" was to attain "a diligent understanding of Indigenous social life and knowledge of their language" (Nahuelpan 2016, 76). Undoubtedly, for the success of their colonizing endeavor, they had to practice devoted forms of attentive listening to effectively learn the Native language and, through it, to capture the nuances of Mapuche lifeways.

When exercised from within a critical and self-reflexive Indigenous framework, however, allkütun can and does interfere with the auditory and sonic scripts of colonialism. It also acknowledges the voices, sounds, and conversations that underlie the acoustics of the larger Mapuche and Indigenous movements of resistance and liberation. In this sense, allkütun echoes what Robinson calls "critical listening positionalities," that is, a critical "filter" in the auditory field that leads to "self-reflexive questioning" against oppressive power relations and opens space for "counter normative" and "resurgent listening practices based in forms of Indigenous sensory engagement and ontologies" (2020, 10–11).[24] Allkütun underlies and traverses this book as a Mapuche auditory positioning and methodological principle that invites us, constantly and critically, to sharpen our ears and modes of listening as part of an anticolonial liberation movement in the sensory, perceptual, and historical realms.

In this critical endeavor, allkütun has enriched my scholarly journey as an engaged way of listening to literary texts, radio shows, audiovisual works, and musical registers from Ngulu Mapu. Of course, the term *allkütun* has formal semantic equivalencies in other languages and contexts. Indeed, in Spanish, the first meaning assigned to the verb *escuchar* (to listen) is defined as "prestar atención a lo que se oye," to pay attention to what is heard (Real Academia de la Lengua Española 2001, 964). In Portuguese, *escutar* is similarly characterized as "tornar-se ou estar atento para ouvir; dar ouvidos a"—a definition amplified through a second meaning: "Aplicar o ouvido com atenção para perceber ou ouvir" (Buarque de Holanda Ferreira 1980, 705). In English, the verb *to listen* is also defined in similar terms. To listen is "to hear attentively; pay attention to (a person speaking or some utterance)" (Oxford University Press 2007, 1616). Likewise, in French, the verbs *écouter* is defined as "to hear with attention," thus forming part of the conceptual field associating the human ear with the mind.[25] In his analysis of the terms *écouter, ouir,* and *entendre* within this same semantic field, philosopher Jean-Luc Nancy attributes a higher level of complexity to the last of these verbs, since with *entendre* the act of listening is interwoven with a deeper level of understanding. *Entendre* invokes both the sensorial and the cognitive (Nancy [2002] 2007, 5–6). Each language has its richness and fosters multiple relationships and conversations. Given Mapudungun's self-definition as language, sound, voice, and territorial expression, allkütun entails a sense of aural attention that directs us uniquely to the environment and, in a deeper ontological and epistemological sense, to the Mapu. Allkütun enlivens the critical, imaginative, and interpretive "ears" of this study, as I engage written, visual, oral, and musical texts, treating them as surfaces and membranes of language permeated by resonant sounds and histories.

Driven by this methodological impulse, this book also historicizes the voices, sounds, and modes of listening that reverberate in language and media. Sound is where the Mapuche linguistic, cultural, racial, gendered, material, and symbolic experiences that exist under, against, and beyond colonialism take shape through form, language, and semantics. The literary, artistic, and cultural productions discussed in this book emerge from the historical formation of Chilean colonial capitalism and the life experiences of Ngulu Mapu as a colonized Indigenous territory. Therefore, history plays a critical role in my analysis of the narratives of both disfigurement and acoustic colonialism in Creole settler literature and media, and of Mapuche interferences. As aptly noted in *¡ . . . Escucha, winka . . . !*, "the use of history" is a "first step in the process of decolonization, in which what is our own and what is foreign are established as part of a positioning" (Mariman Quemenado et al. 2006, 261). Given the mobility and porousness of what is "our own" and what is "foreign" (wingka) in the sound sphere, historical contextualization helps situate colonial articulations and expressions of indigeneity within the complex web of relationships that undergird Mapuche life. Historicization, therefore, involves methodological as well as political commitment.

FORMS OF MAPUCHE INTERFERENCE

For the Mapuche, this soundscape is a battleground where they struggle to undermine, disrupt, and rechannel the continuous waves of colonialism. In their long history of creative agency, Mapuche writers, artists, and media producers in the late twentieth and early twenty-first centuries have used sound strategically. They have oscillated between genres and codes rooted in ancestral traditions, on the one hand, and the imaginative use of technologies and forms of representation from non-Indigenous hegemonic cultures, on the other. These Mapuche creative practices in the sound sphere are hardly the expression of ahistorical "difference" or "otherness," as settler scholars tend to posit in accordance with their own fictional ideations of indigeneity. Rather, these Mapuche artistic and media practices tend to be situated within contemporary struggles for self-representation and agency that challenge existing power relations in Chilean settler society.

In taking a historically situated, relational approach, my study offers a critical route to the audible appreciation of the multiple strategies that Mapuche writers, artists, and media activists utilize to warp voices and sounds in the performative terrains of literary writing, radio, and music. I approach these strategies as intentional *interference* that disrupts the regime of acoustic colonialism. This interference underscores the possibilities for Mapuche agency in their aesthetic, cultural, political, and media endeavors. In its sonic sense, interference denotes the "disturbance of the transmission or reception of radio waves by extraneous signals or phenomena" (Oxford University Press 2007, 1409). In the domains of language, representation, voices, and sound, the agency of the Mapuche disrupts colonial normalcy. The notion of interference captures the aesthetic, political, and historically rebellious impulse of Indigenous creative works and initiatives.

That said, the very concept of interference suggests its limitations. Although literary, artistic, and media interventions may tease out, annoy, interrogate, and disturb acoustic colonialism, they do not dismantle colonial domination. They do not replace other social and political actions that aim to decolonize and liberate. By considering the creative labor of writers, artists, and media producers within a broader set of oppressive structures and systems, this book seeks to avoid the overinflation of the field of "culture." Hence, the chapters first examine hegemonic mediations over the sounds of Ngulu Mapu before analyzing Mapuche interference.

Chapter 1 demonstrates how listening, sound, and colonialism became interwoven in Chilean writings around the time of the "Campaign of the Pacification of Araucanía." It focuses on the brief novel *Mariluán* (1862), by the canonical Chilean author Alberto Blest Gana. My reading of this literary text documents how Blest Gana crafted his portrayal of the Mapuche through the distorting filter of the omnipresent narrator's colonial ear. I demonstrate how the settler body becomes the perceptual and representational border between Chilean society and Mapuche territory as it exercises a misrepresentation of Native voices and sounds.

Through the mediation of his colonial ear, Blest Gana's narrative prose projects a politics of acoustic disfigurement that will prevail as a way to (mis)represent the Mapuche and Ngulu Mapu in the hegemonic Chilean mediascape.

One hundred years later, acoustic colonialism continues and persists. Nevertheless, between the late nineteenth and mid-twentieth centuries, mediascapes have changed dramatically. Throughout the first half of the twentieth century, writing-based media (literature, newspapers) coexisted with another sonically powerful communication technology, namely, radio. Chapter 2 discusses the radiophonic and musical impersonation of a "Mapuche" by a popular character called Indio Pije (Indian Snob). The Chilean comedian and actor Ernesto Ruiz originally staged and voiced the Indio Pije on *Residencial La Pichanga*, a *radioteatro de humor* (comic radio show, or comedy radio theater) that started around the mid-1950s and it reached the height of its success on radio in the 1960s and 1970s. *Indio Pije* also appeared as a character in a *revista de historietas* (comic magazine) launched in 1965, edited and written by César Enrique Rossel, the creator and director of *Residencial La Pichanga*. In 1975, Ruiz issued a musical album titled *Qué pasa en la ruca: Show de cumbias* (What's Happening in the Ruca: Cumbia Show), with six cumbia songs performed by the Indio Pije. I argue that the supplanting of Indigenous voice and body by these radiophonic, musical, and graphic impersonations of the Mapuche is a continuation of the Chilean colonial politics of mediation and distortion. Together, chapters 1 and 2 offer a critique of the practices of listening and acoustic disfigurement of the Mapuche that underlie acoustic colonialism. Thus, in these opening chapters, my own exercise of *allkütun* as a "critical listening positionality" (Robinson 2020, 2) already constitutes an act of Mapuche interference.

I then turn to examining Mapuche interferences that have made use of different media, writing, radio, and music, to disrupt the prevalence of acoustic colonialism during the era of neoliberal expansion. Language was the starting point of this sonic activism. Ever since the establishment of the Chilean nation-state and the colonization of Ngulu Mapu, the sounds of Mapudungun have produced dissonance in the imposed prevalence of the Spanish language. Mapuche poets of the 1980s used Mapudungun in their creative works to redraw the linguistic ecology of Chilean literature. By making the native language audible in their writing and poetry recitals, authors like Leonel Lienlaf, Elicura Chihuailaf, María Teresa Panchillo, and Lorenzo Aillapan Cayuleo interfered in the

Spanish monolingualism that had long dominated the country's literature. These interventions were not simple exercises in linguistic alterity; the literary and aesthetic use of the Mapuche language signaled an ontological and epistemological turn. The meaning of *Mapudungun*, after all, results from the coupling of the terms *Mapu* (land, territory, earth, and universe) and *dungun* (language, sense, sound, and voice). Mapudungun is the interweaving of language, sound, and voice with the land and the environment. By positioning human and nonhuman voices and spaces as a code of communication and representation, Mapudungun challenges anthropocentric definitions of language. It becomes the language of the land, the earth, and the universe. The use of Mapudungun in literary registers, along with Mapuche political and cultural activism, thus provides a multilayered engagement with the sounds and lives of Ngulu Mapu, a distinctive push against acoustic colonialism.

In this vein, chapter 3 delves into the imaginaries of voice, sound, and listening in the poetry of native speakers who work in both Mapudungun and Spanish. In 1989, Leonel Lienlaf published his first collection of poems, *Se ha despertado el ave de mi corazón*, and in 2003, he followed with a second collection, *Pewma dungu / Palabras soñadas*. The Mapuche nexus of language and territory offers a poetic discourse marked by Spanish and Chilean colonial histories, the vital and symbolic force of dreams (*pu pewma*), and the poetic persona's desire for the emancipatory restitution of the Mapuche chant (*ül*). Also in 2003, Lorenzo Aillapan Cayuleo published *Uñumche*, a poetic portrayal of the native birds. Through the bird-person figure of the *uñumche* and avian onomatopoeias, Aillapan Cayuleo highlights the nonhuman sonorities that constitute the language and music of his territory of origin. Chapter 3 consists of close readings of, and listening to, the poetry of Lienlaf and Aillapan Cayuleo as verbal, vocal, and corporeal arts that map, affirm, and vindicate the tears and dreams of Ngulu Mapu, embodying forms of poetic interference within a violated linguistic and acoustic ecology.

Chapter 4 further explores the role of Mapuche linguistic, musical, and political vocality in rebuilding the public audibility of Ngulu Mapu. It examines radio as a platform for an Indigenous interference with the hegemonic waves of acoustic colonialism. I specifically discuss the experience of *Wixage Anai*, a radio program that began broadcasting in June 1993 in Santiago. The program reached audiences throughout Ngulu Mapu and even across the Chile-Argentina colonial border that divides Wallmapu. Produced and directed by Mapuche activists, *Wixage Anai* alternated between Mapudungun and Spanish. It was a collective, volunteer initiative determined to put the Mapuche people's language, culture, and political struggles on the airwaves. In the postdictatorship setting of the 1990s, *Wixage Anai* engaged with the resurgent movement to recover Mapuche lands. Based on a Mapuche poetics and politics of communication centered on *nütram* and *allkütun,* this use of radio brought linguistic, musical, and cultural endeavors together with political activism and enabled the Mapuche to build public audibility and agency.

Finally, chapter 5 focuses on the music of contemporary Mapuche life. First, I examine the influence of *los Mexicanos,* a type of radio show broadcast prolifically by radio stations of Ngulu Mapu that, with the broad mass impact it sustained from the 1960s onward, connected Mapuche listeners across the rural/urban divide. The chapter also discusses the work of the Mapuche group Wechekeche Ñi Trawün and their creative practice of rap music, a genre that has gained popularity among a sizable cohort of the late twentieth- and early twentieth-century new generation of Mapuche for whom music constitutes, among other things, a political tool. I end this final section of the book with a discussion of the continuity of *ül,* Mapuche chant, focusing on the work of Elisa Avendaño Curaqueo, an *ülkantuchefe* (chanter) who has cultivated the genre and has conducted important research on it. In short, chapter 5 is an exercise in attentive listening to the heterogeneous musicality that shapes the sonic environment of contemporary Mapuche life, in which multiples musical registers overlap as part of an aural and temporal continuum in Ngulu Mapu.

Guided by the Mapuche concept of allkütun and a sense of historical engagement, in this book I exercise literal and figurative modes of listening. As a comprehensive approach, "listening attentively" enriches the possibilities of analysis and interpretation of what I define as Mapuche (or Indigenous) *interference.* It is a form of listening that strengthens and interweaves close reading, aural attention, and historical contextualization. My critical journey thus engages with the poetics and the sonorities of texts to delve into their multiple historical and imaginary relationships and resonances.

ONE

.....

Disfiguring and Silencing of the Mapuche in the 1860s

Like drawing, painting, and photography, written text in genres ranging from history books and literary works to newspapers was an enduring technology of representation and mediation of Native life in Ngulu Mapu. From early childhood, I was exposed to the presence of the wingka lifru *(settler book). The rural Mapuche-Williche community in which I grew up—Tralcao—was subjected to Catholic evangelization, Western education, and the Chilean state's citizenship regime since the late nineteenth century. In the home of my grandmother, María Clara Lefno Huechante, who raised me, the few books available were mostly Christian sources. Resting on my grandmother's altar were a 1961 Spanish edition of the Bible, an edition of the Old Testament, a book on the Catholic saints' lives, and another one of prayers. Elsewhere in our old wooden house, I remember a couple of books on history and botany. In my elementary school, located in Tralcao's Catholic chapel, where most of my classmates were from Mapuche-Williche families, the few books available were children's literature and official textbooks, all of them by Chilean, European, and American authors. At home and at school, we only had access to a limited Chilean library. Then, while attending high school in the city of*

Valdivia between 1976 and 1979, where my journey through the lettered culture of the "Republic of Chile" continued, my education arrived at the writings of Alberto Blest Gana, who by the early twentieth century had become required reading in the Chilean school system. His novels Martín Rivas *(1862) and* El loco estero: Recuerdos de la niñez *(1909), both set in Santiago, were officially adopted in the Chilean high school curriculum. Blest Gana had long been a canonical reference in the literary imaginary of Chile as a settler nation. Indeed,* Martín Rivas *is traditionally considered the inaugural "Chilean novel," which like* Mariluán*—the brief novel I discuss in this chapter—was published the same year that the military invasion of Ngulu Mapu began. With the tools the Mapuche political and cultural movement has given me, in addition to the teachings of my kuifikeche (Mapuche elders), my critical reading and listening of Blest Gana's novel* Mariluán *aim to contribute to the communal saga of disassembling and interfering with the orderliness of the colonial library.*

....

At the center of Manuel José Olascoaga's pictorial portrayal of the encounter between Chilean army officials and Mapuche authorities in Hipinco in 1869, Colonel Cornelio Saavedra sits comfortably on a chair with wide back support, under the shade of a native tree, legs crossed, dressed in a neat military uniform (see figure 1.1). This same Saavedra planned and led a Chilean colonial invasion and occupation of Mapuche territories in the second half of the nineteenth century—a military operation euphemistically known as the "Campaign of Pacification of the Araucanía."[1] Over subsequent decades, especially through the agency of Chilean history books, this painting has been impressed on thousands of eyes in Ngulu Mapu / Chile. It is a visual image that established the wingka as "authority," the one who sets the terms of the conversation, the one in control of colonial mediation. Indeed, Colonel Saavedra is at the center, dominating the image as he dominated the moment, so our eyes are drawn to him first. Only after focusing on Saavedra do our eyes move to a group of Mapuche, some sitting on the ground, others standing, encircling him. Several Mapuche men wear *trarilonkos*, the traditional headbands the Mapuche wear for important events, to signal the political relevance of the gathering. To one side of Saavedra, we can observe men wearing hats of the typically Chilean Creole style of that

FIGURE 1.1 · Photographic copy of Manuel José Olascoaga's painting of the Hipinco Parliament (1869). José Toribio Medina Room, National Library, Santiago, Chile.

time. Standing in the middle, a man, also in characteristic Creole attire, appears to be addressing Colonel Saavedra. Behind him, receding into the painting's blurred background, is a semicircle of men on horseback who seem to be Chilean troops. In a logic of representation that operates in similar ways to the realm of literature, this visual image places not only the gaze but also the ear of the settler at the center.

Olascoaga's painting has had wide circulation over time, being often cited by scholars or included in school textbooks in Chile.[2] The pictorial work has not survived as such but has been preserved and archived in Chile's National Library through an image captured by an unidentified photographer of the time.[3] Olascoaga's painting is a portrait of the Hipinco Parliament (*parlamento*), an assembly that took place on December 24, 1869, in the Hipinco plains, not far from the town of Angol. This *parlamento* brought Saavedra and his men together with Mapuche representatives to negotiate amid the state-sponsored military invasion of Native lands still under Mapuche control. The colonial military enterprise led by Saavedra started in 1862 with the seizure of several towns, such as Mulchen, Negrete, Angol, and Lebu. Subsequently, a wider military deployment entailed the violent removal of many Mapuche communities (Pu Lof) from their territories. Ironically,

Olascoaga's portrayal of the Hipinco Parliament, when used as an illustration in books about the history of Chile, is often accompanied by a caption that labels it a *reunión amistosa* (friendly meeting), with no further commentary on the fact that this type of negotiation took place under military pressure, and was itself a type of colonial invasion, occupation, and settlement.[4]

What is striking in this pictorial representation is how the very corporeal and spatial position of the Chilean military chief reveal the moment's actual asymmetrical power structure for viewing, speaking, and listening. Many of the Native attendees are Mapuche authorities, yet for Saavedra they are "seen" and "heard" as mere objects of geopolitical subordination: as either "friendly Indians" or "cannon fodder" in the ongoing Chilean war of "pacification." Whether the Mapuche are submissive or not, they will be considered "objects of," as symbolized in the image of the Hipinco Parliament. From his podium of sorts, supported by a surrounding crowd of non-Indigenous subalterns, Colonel Saavedra clearly mediates the conversation. His body and ears constitute a sort of *frontera* (border). From the center, he hierarchically filters the voices and silences of the Indigenous representatives. His body's position and posture stage the unequal power relations in the realm of communications between Chilean settlers and the Mapuche. This colonial authority-centered setting is symbolic of how Chilean state agents and settlers would interact with Mapuche authorities across time in their efforts to affirm a tutelary mediation in the process of listening and speaking.

This well-known visualization of aural mediation also reflects how colonial and racialized modes of listening and representing the Mapuche resonate in the Chilean lettered archive. To illustrate these representational practices, this chapter focuses on how Mapuche characters and sounds are both figured and disfigured through *Mariluán*, a novel written by Chilean Creole writer Alberto Blest Gana, a central figure in the formation of the Chilean literary canon. The novel was published in Santiago in 1862, right at the beginning of the infamous decade of the Chilean military invasion and subsequent cultural occupation of Ngulu Mapu led by Colonel Saavedra, which haunts and beleaguers us to this day. Blest Gana's novel can be read as part of what Congolese scholar V. Y. Mudimbe designates as the "colonial library," a concept he based on the experience of Africanism and the discursive production of an "idea of Africa." He defines a "colonial library" as "a body of knowledge constructed with the explicit purpose of faithfully translating and deciphering the

African object," which would unveil "its being, its secrets, and its potential to a master who could, finally, domesticate it" (Mudimbe 1994, xii). I read Blest Gana's novel as part of the Chilean colonial library and, more specifically, as a literary work that aims to construct an Araucanian object to be domesticated in the sphere of discourse and representation. In its hegemonic eagerness to institute an "idea of" the Mapuche and apprehend its indigeneity, Blest Gana's novel can also be read and "heard" as a discursive resonance of the Pacification Campaign.

My analysis of the novel *Mariluán* interweaves textual analysis and historical contextualization with the Mapuche listening principle of allkütun, which enables me to exercise close reading *as* close listening in a literary, figurative sense. This approach to Blest Gana's novel allows me to unveil how a colonial ear constantly filters and regulates the voices and sounds of the Mapuche, establishing the limits between the colonizing and colonized societies. At a historically decisive moment of the Chilean settler occupation of Ngulu Mapu, Blest Gana's narrative operates as a colonial ear and pen vis-à-vis the Chilean-Mapuche frontier, while it also reveals the failures of assimilationist colonial desire and the delimited "civilizational" framework imposed on Native bodies, voices, and lands in the consolidation of the nineteenth-century Chilean nation-state.

On the other side of the cordillera, in the Mapuche Lands of the East (Puel Mapu, or what is now called Argentina), the Western trope of "civilization" was already at the core of the literary script of Argentine settler colonial writers of the period. The so-called Conquest of the Desert (1879–83) in Puel Mapu is a historical parallel to the Pacification Campaign in Ngulu Mapu, even though it was a more devastating genocidal enterprise on Mapuche and Indigenous lands.[5] There, Domingo Faustino Sarmiento, a literary icon of the Argentine Creole elite, writes, "Para nosotros, Colocolo, Lautaro y Caupolicán, no obstante los ropajes nobles y civilizados con que los revistiera Ercilla, no son más que unos indios asquerosos, a quienes habríamos hecho colgar ahora" (For us, Colocolo, Lautaro and Caupolicán, even though Ercilla dressed them in civilized and noble apparel, they are simply filthy Indians whom we should have executed by now).[6] Argentine writer and scholar David Viñas argues that this type of racist and genocidal statement is a projection of Sarmiento's "popular education program" that "appealed to the 'whitening' or replacement of the Indian (as well as of the gaucho) through European immigration" ([1982] 2003, 63). Indeed, the literary intelligentsia of the Argentine settler colonial nation-state aimed to eliminate the Indian from their liberal, "civilizational" imaginary, following Sarmiento's path. As a dissident variant, other writers, such as José Hernández in his epic poem *Martín Fierro* ([1872; 1879] 2001), include the popular and rural "voice" of the gaucho as the *hombre argentino*, exalting it through the metanarrative of what scholar Josefina Ludmer characterized as the "gauchesca genre" ([1988] 2000).[7] As a "popular" Creole or mestizo figure in terms of race, the gaucho represented a less threatening and more assimilable figure for the "civilizational" nation-state.

The Chilean lettered elite were not very different from their Argentine peers. The Western colonial trope of "civilization," as opposed to Indigenous "primitivism" or "barbarism," was already part of the foundational pages of the Chilean nation-state's print culture. Indeed, an 1812 editorial in *Aurora de Chile*, the first newspaper of the newly founded republic, states this aspiration: "Los fuertes habitantes de los quartro UltraMapus, los Indios nos prometen una colaboración activa para repeler los insultos estrangeros. . . . Tálvez nos dista el bienhadado momento de su convercion, civilización, y cultura" (The strong inhabitants of the Four Directions, the Indians, promise us an active collaboration in repelling the foreign insults. . . . Maybe we are still far from the blessed moment of their conversion, civilization, and culture) (Henríquez 1812, 2).[8]

Entangled in this broader colonial civilizational project, Blest Gana's novel acts as a filter that continuously regulates the scope and limits of acoustic representation and aurality in the encounter and clash between Chilean Creole society and the Mapuche. The omniscient narrator intermingles as a colonial ear by constantly mediating the narration to figure and disfigure the Indigenous voices and sounds of Ngulu Mapu. This sort of aural filter in Blest Gana's novel works much like what scholar Jennifer Lynn Stoever defines as "the sonic color line," a codifying mechanism that "enables listeners to construct and discern racial identities based on voices, sounds, and particular soundscapes" and, in the process, "produces, codes, and polices racial difference through the ear" (2016, 11). In her study of sound and race in the White and Black American contexts, she demonstrates how this aural codification operates through a "listening ear" that constitutes "a socially constructed ideological system producing but also regulating ideas about sound" (Stoever 2016, 13). The colonial ear also has these ideological and racial dimensions, particularly in terms of how the White Creole self acoustically filters and codifies, figures and disfigures, the sounds of the Native.

Ana María Ochoa Gautier has discussed the role of "the ethnographic ear" in the linguistic, musical, religious, and civic orchestration of the nation-state's policies and actions. Ochoa Gautier specifically analyzes the "poetics and politics of hearing" as staged in ideological clashes between conservative and liberal intellectuals in late nineteenth-century Colombia. In a historical period in which the ear acquired a "central role in knowledge constitution" across different disciplines (i.e., linguistics, medicine, and psychoanalysis), it was critical for the Creole lettered elite (men) to "control the processes of language purification and their relation to power enacted a simultaneous purification of the practices of the ear" (Ochoa Gautier 2014, 163). Here writing emerges as a disciplinary regulator intertwined with the regimes of the ear and orality, which Ochoa Gautier describes as a process of acoustic and vocal immunization: "Vocal immunity uses the fear of voice's intrinsic potential to manifest an incoherent or otherwise undesirable form of the self to produce a vocally articulate one, grammaticalizes the voice through the rules of writing while purporting to speak in the name of 'people's' audible vocality, and curtails the dubious ear's reception of the voice by training it to distinguish and parcel out the uses and functions of proper and improper voices amongst different peoples" (2014, 171). Ochoa Gautier unveils an interdependence between the disciplinary functioning of the ear, orality, and writing in nineteenth-century Colombia, where philology came to occupy an influential status.

Ochoa Gautier's description coincides with the modes in which Chilean lettered men and literary discourses put into action mechanisms that narratologically regulated the circulation of "proper and improper voices" and articulated the "vocal immunity" of the language and culture of a Creole "republic" in search of its consolidation. My approach situates the post-1800 "republic" in Latin America as a nation-state project entwined in a process of colonial settlement (Cárcamo-Huechante 2023; Speed 2017), which structurally and historically acquires shape and content through the systemic subjugation and racialization of Indigenous peoples and the consequent occupation and dispossession of their lands. In this saga, the regimentation of listening and the voice is based on the "immunizing" preeminence and omniscience (Ochoa Gautier 2014, 171) of what I call the colonial ear. This ear becomes key to a colonizing mechanics of mediation, projection, substitution, and erasure that maintain control over the sonic and aural domain of representation on behalf of the Chilean settler nation-state project and its aim to eliminate or subordinate the Native. Thus, this practice of hearing and filtering forms part of the vast and enduring deployment of *acoustic colonialism*, a regime that establishes modes of listening, voicing, sounding, and silencing that reinforce the oscillating colonial and racial politics of exclusion and subjugated inclusion of Indigenous peoples, as the Chilean nation-state has practiced throughout history.

PRELUDE

Alberto Blest Gana's brief melodramatic novel *Mariluán* was originally published in Santiago in 1862, in the newspaper *La Voz de Chile*, whose name establishes the "voice" of the settler nation as a metanarrative.[9] The plot of *Mariluán* unfolds around the experiences of the young Mapuche Fermín Mariluán, a leading officer in the Chilean army, as he returns from Santiago to Ngulu Mapu.[10] Mariluán arrives in the border city of Los Angeles in April 1833. Historians have noted that Blest Gana based the character on the son, also named Fermín, of the prominent Mapuche chief (longko) Juan Francisco Mariluán (Pinto Rodríguez 2003, 259).[11] In giving the character Fermín's genealogy, Blest Gana constructs him as a figure occupying a liminal, ambivalent space between fiction and history. This crossing between the historic and the fictive relates to Blest Gana's aesthetic ideology as a writer, characterized by his hybrid amalgam of Romanticism, naturalism, and realism, even though he preferred the latter. On June 24, 1856, he wrote to his friend José Antonio Donoso that "prefiero las novelas de estudio social" (Blest Gana 2011, 19). Without discarding other trends, this statement reveals his personal inclination to the tradition of literary realism.[12]

When *Mariluán* begins, Fermín has gained a high rank in the Chilean army, which he had entered after completing his secondary studies at the Lyceum of Chile in the capital city of Santiago. Despite having made his reputation as a "brilliant cavalry officer" in the Chilean military, Fermín returns home with the firm intention of reconnecting with his people and organizing an uprising against Chilean colonial domination. The story becomes unexpectedly complicated when the protagonist falls in love with Rosa Tudela, a young woman from a local Creole family. Since the death of her father, Rosa has been under the tutelage of her brother Mariano and her mother Andrea Ramadillo. Her romance with Fermín threatens family plans, especially for the patriarchal and overbearing figure of Mariano. The Tudelas thus conspire to arrange a marriage between Rosa and a prosperous businessman from Talcahuano, don Claudio Retamo (Blest Gana [1862] 2005, 17–18).[13]

Love, business, and politics are recurrent themes in Blest Gana's prose (Concha 2011, 47; Hosiasson 2017, 240; Sommer 1991, 204–20). Sentimental plots are interwoven with matters of money and business, just as romance mixes with political intrigue in nineteenth-century Chile. Also published in 1862, Blest Gana's *Martín Rivas*, over three hundred pages long, earned the highest position in the Chilean literary canon as a "national novel," which put *Mariluán* in a relatively marginal position (Láscar 2003). While *Mariluán* addresses the relationship between Chilean and Mapuche societies as a terrain of turbulence and crisis, *Martín Rivas* allegorizes the class and family alliances within an ethnically and racially Creole society. The latter tells us the story of Martín Rivas, a young Creole liberal who moves to Santiago from a northern province in search of social and economic success. He falls in love with the daughter of a conservative oligarchic family. In its allegory of liberal-conservative coexistence, hegemonic class alliances, and the imaginary of the Creole family within the narrow geography of Chile's Central Valley, the novel excludes questions of indigeneity and racial difference in its "imagined community" (Anderson 1983). *Martín Rivas* thus offers an indulgent image of identity for the limited circle of Chilean readers: The educated Creole minority settled in the Chilean Central Valley and formed within Eurocentric cultural patterns, self-satisfied by its capitalist and capital-city horizons.[14]

In contrast to *Martín Rivas*, *Mariluán* begins with movement southward, toward a region known in nineteenth-century Chilean geopolitical discourse as La Frontera, the Chilean-Mapuche frontier. Mariluán's movement removes Blest Gana from the central geographical axis (i.e., Santiago, central Chile) around which the narratives of his literary works often revolve. *Mariluán* appears to amplify the author's settler Chilean "national" desire for geographical expansion. In an era of capitalist "expansion" in Chile, Blest Gana's novelistic journey reveals a geopolitical eagerness to assimilate the Mapuche territories into the hegemonic national imaginary.[15] Indeed, by the mid-nineteenth century, the possession of Ngulu Mapu lands had gained strategic relevance for the extractive and export-oriented economics of Chilean capitalism, which had already succeeded in establishing the mining industry in the north. The Chilean occupation of Mapuche territories in the 1860s was concomitant with the expansion of agriculture and the cattle industry, driven by the new domestic and international demands (Vitale 2011, 304–8) as well as the emerging hegemonic liberal vision of progress and modernization in Chile (Sepúlveda 2018, 179–94). In this entanglement between expansive geographies of capitalization and colonization, the literary cartography of Blest Gana also had to be updated.

The novelistic turn toward La Frontera exposed Blest Gana's authorial and narrative persona to ethnic, racial, and cultural difference, straining his narrow Creole horizons. *Mariluán* institutes a literary economy marked by a racialized novelistic mapping and possessing of the human geography of Ngulu Mapu. Additionally, the novel is a test for state geopolitics and the cultural project of assimilation of the Mapuche. In his novelistic journey, the literary persona of Blest Gana—as narrator and author—relates to the Mapuche environment exclusively from within the boundaries of Los Angeles, a well-shielded urban, administrative, and military Creole settlement, from which he never ventured farther into La Frontera. From there, at a certain distance, his narration "represents" the Mapuche. His constructions of images and voices around Ngulu Mapu are mediated by what Blest Gana may have heard—knowledge *from hearsay*. Orality and listening underlay the very weaving of *Mariluán* as a written artifact as well as a story. The narration of *Mariluán* abounds in the comings and goings of characters who act as messengers from the inlands of Ngulu Mapu. The Mapuche territory is portrayed as an exogenous, distant space, as an *outside* in distinction to the colonial enclave of Los Angeles, a city positioned as the axis of reference in the novel's geography.

Considering these aural and spatial entanglements, I approach *Mariluán* as a literary mechanism of mediation and disfigurement in listening, voice, and sound. My reading diverges from the thesis posited by historian Jorge Pinto Rodríguez, who, from a Chilean viewpoint, asserts that Blest Gana's novel constitutes part of a discourse in which "our novelists and poets join their voices to those of the Mapuche" (2003, 271). I argue that the story symbolizes rather the dispossession of voice and the silencing of the Mapuche; indeed, the Indigenous protagonist of the novel ends up both literally and allegorically headless. In this sense, Blest Gana's novel ultimately succeeds in reinforcing the settler politics of assimilation, disfigurement, and annihilation of the *voices* of the Mapuche.

The first lines of *Mariluán* foreground its Indigenous protagonist's countenance:

> La indómita energía de la raza inmortalizada por los cantos de Ercilla, brillaba en los ojos de Fermín Mariluán. En un pecho espacioso y levantado, latía su activo corazón, cuya viril entereza daba a sus negros y pequeños ojos su tranquilo mirar, y a los labios, algo abultados, la fría expresión de orgullo que caracteriza la fisonomía de los araucanos. (Blest Gana [1862] 2005, 5)
>
> The untamed energy of the race immortalized by the songs of Ercilla sparkled in Fermín Mariluán's eyes. In his proud, broad chest beat his active heart, whose virile integrity gave to his small, black eyes a calm gaze, and to his voluminous lips, the cold expression of pride that characterizes Araucanian physiognomy.

From the very start of the novel, the narrator places the Indigenous figure of Mariluán within a Chilean metanarrative based on representations of the Mapuche forged by Spanish colonial discourse: the archetype of the "Araucanians." The point of reference here is Alonso de Ercilla y Zúñiga's sixteenth-century epic poem *La Araucana*. Ercilla y Zúñiga accompanied conqueror Pedro de Valdivia during the first decades of the Spanish invasion of the Mapuche people's lands, the beginning of the long Mapuche-Spanish confrontation that became known as the Arauco War. As scribe and soldier, Ercilla y Zúñiga wrote *La Araucana* between 1550 and 1563. He offered European readers, then and in subsequent centuries, an epic image of the inhabitants of "Arauco." In Ercilla y Zúñiga's poem, the Mapuche race is strong and courageous in war, comparable to the male archetypes of classic Greek and Latin epics. Through a sort of mirroring effect, this iconic imagery enhanced the self-figuration of the Spanish conqueror and colonizer of Ngulu Mapu in general. *La Araucana* went on to become a foundational text of the "colonial library" (Mudimbe 1994, xii) that circulated in European elite circles during the era of Spanish colonialism and was widely reproduced later, in post-1810 republican Chile.

Blest Gana's is not the only author in Chilean Creole culture to invoke *La Araucana*'s epic imaginary as a representational filter for the Mapuche.

The Chileanization of Ercilla y Zúñiga's Spanish colonial epic poem has an extensive history. Indeed, Chilean nation's foundational writers, such as Bernardo O'Higgins, José Miguel Carrera, and Francisco Antonio Pinto, embraced the archetypes of the Native warriors Ercilla y Zúñiga portrayed to give a symbolic ethos to their struggle against Spanish domination. Later canonical Chilean writers also celebrated *La Araucana* as a foundational text. In 1971, several authors contributed to a collection of essays published by the Catholic University of Chile and tellingly titled *Don Alonso de Ercilla: Inventor de Chile.* Chilean Creole authors praise Ercilla y Zúñiga as the "inventor of Chile." Contributors include Pablo Neruda, who received the Nobel Prize for Literature that year and was an iconic member of the Communist Party of Chile. In the book's opening essay, in reference to Ercilla y Zúñiga, Neruda writes,

> A él le debemos nuestras constelaciones. Nuestras patrias americanas tuvieron descubridor y conquistador. Nosotros tuvimos en Ercilla, además, inventor y liberador. . . . Ercilla no solo vio las estrellas, los montes y las aguas, sino que descubrió, separó y nombró a los hombres. Al nombrarlo, les dio existencia. El silencio de las razas había terminado. (Neruda 1971, 12)[16]

> We owe our constellations to him. Our American homelands had a discoverer and a conquistador. In addition, in Ercilla we had an inventor and a liberator. . . . Ercilla not only saw the stars, the forests and the waters but he also discovered, set apart and named the men. By naming them, he gave them existence. The silence of races had ended.

In praising Ercilla y Zúñiga, Neruda adopts an explicit settler tone and sensibility. Not only does he celebrate the figure of Ercilla y Zúñiga; he also assigns an ontologically and historically constitutive power to colonial writing. In his Eurocentric, lettered view, Native peoples, lacking the Western technology of writing, seem to be relegated to a status of nonexistence: "the silence of races."

For centuries, *La Araucana* has not only become part of the pantheon of Chilean literature but also reverberated in the popular, civic, military, and institutional imaginations of the settler colonial republic since the early nineteenth century.[17] In this regard, what I call the Chileanization of *La Araucana* has been critically examined by literature scholars. For example, poet and essayist Waldo Rojas has discussed the continuous appropriation of Ercilla y Zúñiga's poem in foundational nation-state discourse and in popular culture as well as in the writings of prominent poets and writers in Chile such as Neruda (Rojas 1997). Scholar Bernardo Subercaseaux offers a broader panorama on the interpretative appropriation of *La Araucana*, including a reference to a more recent work on military history, *Historia del Ejército de Chile Tomo I*, published by the Estado Mayor of the Chilean army in 1980, during the Pinochet dictatorship. The latter book's embrace of Ercilla y Zúñiga's epic poem is explicit: "The work of Alonso de Ercilla, *La Araucana*, has been fundamental in this aspect and its stanzas have served as patriotic prayers to uplift the Chilean spirit in difficult moments" (Subercaseaux 2021, 15). These studies make clear that this epic poem constitutes a formative text in the Chilean cultural imaginary.[18] Through a Western-Hispanic lens, Ercilla y Zúñiga's text established "an idea" of "Arauco" and the "Araucanians," while simultaneously paving the way for what later will constitute the foundation of "Chilean literature"—that is, a literature written in Spanish on lands where Spanish was not the first language.

By interweaving its opening lines with Ercilla's archetypical representations of the Native and a Spanish-language literary legacy, Blest Gana follows a colonial lineage of writing and cultural imagination. The (male) desire of the Creole lettered subject thus establishes continuity with his Spanish counterpart. The coupling of literary writing and colonialism carries over from Ercilla into Blest Gana, from Spanish epic imagination to Chilean novelistic discourse. This continuum of the colonial imaginary and the history of writing on the Mapuche leads Blest Gana to reiterate the archetype of the "Araucanian." As a subject, the Araucanian is split between the fierce "savage" and the virtuous and courageous warrior comparable to the epic heroes of *La Araucana* and, more broadly, of Western literature.

In Blest Gana's novel, the narrator highlights Fermín Mariluán's "virile integrity" by evoking the colonial epic construction of the ancient "Araucanians." From a Chilean perspective, these words are complemented by the voice of one of Mariluán's "fellows in the Lyceum of Chile," who describes the protagonist as "a brave enthusiast of military glory" (Blest Gana [1862] 2005, 5). As in *La Araucana*, Blest Gana's narration shows a propensity to profile the Araucanians, all men, as markedly physical, instinctive, and sentimental characters who lack intellectual capabilities. Mariluán's reasoning is limited to military and moral affairs; as he succumbs to emotional urges, "the physiognomy of the Araucanians" becomes central. In framing his Indigenous characters, Blest Gana draws on the rhetoric of physiognomy common in narratological models of naturalist and realist novels of the eighteenth century. The Chilean author aesthetically incorporates this technique in his works, creating literary portrayals rich in physical characterization. Blest Gana also adds a layer of assimilationism by converting Mariluán into a character permeated by colonial military morale and culture:

> Los que le vieron ceñir su espada de oficial recuerdan el garbo con que llevaba el galoneado uniforme de caballería, y el cariño con que colocaba su mano pequeña, herencia de su raza, sobre la empuñadura de esa espada, como impaciente de tener ocasión de sacarla con razón, porque estaba seguro de poder después de envainarla con honor, para cumplir con el lema puesto a las hojas toledanas en palabras como las que hemos subrayado. (Blest Gana [1862] 2005, 5)

> Whoever saw him fasten his officer's sword remembers the elegance with which he carried his decorated cavalry uniform and the care with which he placed his small hand, an inheritance of his race, on the sword's handle, as if he were impatient to have *a reason* to unsheathe it, certain that upon returning it with *honor* he would fulfill the motto inscribed on the blade of Toledo steel which we have emphasized.

The narrator extols Mariluán's artful mastery of the sword along with his adherence to Spanish principles of military discipline and protocol. By highlighting the mottos of "reason" and "honor" inscribed on the blades of a sword made in Toledo, Spain, Blest Gana overlays the Native character and body with a material and discursive cloak of colonial genealogy. Mariluán's very clothes—"the decorated cavalry uniform"—portrays the Indigenous subject as assimilated into the robes of the Hispanic Creole military culture, which differentiates him from the half-naked Indian portrayed by Ercilla y Zúñiga. Thanks to his literacy, Mariluán's "favorite reading" happens to be *La Araucana*. The nineteenth-century Chilean writer thus constructs his own "civilized" Araucanian: "El poema de don Alonso de Ercilla y Zúñiga despertaba en el alma de este indio, pulido por la civilización, ese orgullo que las razas perseguidas cultivan como una religión salvadora" (Don Alonso de Ercilla y Zúñiga's poem awakes in the soul of that *Indian, polished by civilization*, the pride that persecuted races cultivate like a religion of salvation) (Blest Gana [1862] 2005, 7; my emphasis).

What is striking is that the figuration (or rather disfigurement) of the Indigenous character as an "Indian polished by civilization" reveals a construction that, despite subtle variations and disruptions, is plotted through an entanglement of accumulation, juxtaposition, and continuities of colonial representations and mediations. The emergence of "Chilean literature," a process of canon formation to which Blest Gana's literary works belong, does not break with colonial forms of narrativization of the Mapuche. Indeed, *Mariluán* is a contradictory amalgam through which the Spanish epic archetype of the "Araucanian" wild warrior is recast in new molds of indigeneity that emerged in the mid-nineteenth-century Chilean Creole elite's mentality. This poses a dilemma for the Mapuche: One can either become a "patriotic" soldier (emblematic of inclusion, assimilation) or relapse into a "savage" (emblematic of exclusion, elimination). Therefore, *Drama en el campo* (Drama in the Countryside) is a very telling title for the literary trilogy that the novel *Mariluán* forms part of. Blest Gana's narrative works as a "drama" to depict the life of the protagonist. It mixes epic elements with Romanticism, realism, and naturalism inherited from the eighteenth-century novel. In this juxtaposition of epic and novelistic strategies, *Mariluán* is not only *Araucanized*, in Ercilla y Zúñiga's sense, but also *Creolized* in keeping with the racial, ideological, and cultural logic of the emerging Chilean state imaginary of indigeneity, a hegemonic logic that the Santiago-based author of *Mariluán* closely follows.

From the beginning of the novel, the narrator praises Mariluán for excelling in Chilean military feats: "Many soldiers remember having seen him in the Lircay battlefield going further ahead of the rest to confront the enemy on his own, with his eyes radiant with happiness after each cannon stampede" (Muchos soldados recuerdan haberle visto en Lircay adelantarse solo a desafiar al enemigo, con los ojos radiantes de alegría a cada estampido del cañón) (Blest Gana [1862] 2005, 6). Ercilla y Zúñiga's archetype of the brave warrior is now refashioned into a character fascinated by the acoustics of the Western military apparatus. The aestheticized sounds of war permeate the narration and the persona of Mariluán: "He almost liked the music of the big mouths of fire more than that of the harp or the vihuela instrument with which he used to sing" (casi más le gustaba la música de las grandes bocas de fuego que la del arpa o de la vihuela en que solía cantar) (Blest Gana [1862] 2005, 6).

Sound plays a significant role in the staging of the character, language, and narrative scenarios. If the noise of firearms has delighted Mariluán's ears, Blest Gana also exalts his taste in singing and playing Creole musical instruments, such as the harp and vihuela, to show the Native's immersion in wingka society. The "Indian, polished by civilization," pleases the colonial ear with his command of non-Native musical chords. His neat calligraphy further proves his ability to manage the master society's tools: "La misma mano que blandía la espada como un rayo exterminador, que pulsaba las cuerdas de la guitarra para acompañar las alegres canciones que Mariluán gustaba entonar, tenía también el don de trazar una letra elegante, que desde temprano usó para ayudar a los amigos" (The same hand that wielded the sword like an exterminating bolt of lightning and strummed the guitar to the happy songs he liked to sing, could also write elegant script, which he used early on to help his friends) (Blest Gana [1862] 2005, 6).

These qualities make Mariluán into a Chileanized Indigenous persona, legible to an author like Blest Gana from the White Creole elite from Santiago. Two aspects here on authorship and context are noteworthy. First, the representation of Mariluán, based on his mastery of military, musical, and literary skills, is a projection of Blest Gana's self-image. Indeed, in addition to his agricultural background, his family was linked to the military and lettered elite in nineteenth-century Chilean society. Literary critic Jaime Concha highlights that "some maternal relatives" of the author who participated in "the struggles for Independence" may have "predominated in his early formation." After he completed secondary school in the National Institute of Santiago in 1841, he enrolled in the Chilean Military School in 1843 (Concha 2011, 27). Likewise, his brother Guillermo was a poet and playwright, and his brother Joaquín devoted his life to Chilean politics. The Blest Gana family was a portrait of the Creole elite and its characteristic activities: military academy, the letters and the arts, and politics. The figure of Mariluán, with his military and musical skills and his "elegant script," mirrors the very subject of Creole high society and illustrates Blest Gana's idea of a Chileanized "Araucanian."

Blest Gana's Indigenous protagonist thus enters the romance of the Chilean novel as an "Indian polished by civilization." Falling in love with Rosa Tudela, a White Creole woman from Los Angeles, requires

Mariluán to show off his "civilized" virtues to overcome the barriers between the colonizing Chilean society and those who come from Ngulu Mapu. He relies on his "polished" personality to get invited to a party in the Tudelas' social circle and thus to pave the way for his love saga:

> Mariluán contaba con algunos recursos muy importantes para llevar a cabo esta idea que podia abrirle las puertas de la casa en que debía encontrar a su querida: gozaba, por su carácter jovial, de gran popularidad entre las más encopetadas familias de Los Angeles, *tocaba con destreza la guitarra y poseía una voz agradable.* Estas dos últimas cualidades le evitaban la molestia de buscar alguna persona que ejecutase *la parte vocal e instrumental,* que forma el requisito más importante de un esquinazo. (Blest Gana [1862] 2005, 24; my emphasis)
>
> Mariluán had a few resources that could help him open the doors to the house where he would find his love: he had a jovial character, which made him popular among the poshest families of Los Angeles, he played the guitar artfully, and he possessed a pleasant voice. These latter two qualities saved him the bother of having to find someone to perform the *vocal and instrumental* parts necessary for an esquinazo.

To the credentials of his "jovial" Native character, which ensures his popularity "among the poshest families of Los Angeles," the narrator adds "a pleasant voice." Voice becomes a crucial mechanism of social mobility. As philosopher Mladen Dolar notes, "We are social beings by the voice and through the voice," which "stands at the axis of our social bonds" (2006, 14). Mediated by Western philosophy, from Aristotle to Althusser and Agamben, Dolar tends to abstract "the voice" from its corporeality by dismissing its sensory affects and effects. Although he points out that "voices are the very texture of the social" (Dolar 2006, 14), his elaborations become elusive before the sentient "texture" at play in the transition from "the mere voice" to speech or, in a more Western philosophical lexicon, from *phono* to *logos* (105–7). What stands out in Mariluán is that the sensorial quality of his "pleasant voice" functions in unison with the more ideological filters of class, race, and culture implied in the text above. The economics and politics of social climbing that so fascinated Blest Gana here acquire racial dimensions;

moreover, they function through the seductiveness of "the mere voice" along with the manual and corporeal virtues of the Indigenous character, making it palatable to the colonial ear.

With these "vocal and instrumental" skills, Mariluán shows self-confidence in his ability to carry out, as expected, a Creole *esquinazo* performance. In the Chilean musical and social tradition, the esquinazo is a song of salutation to pay homage to someone, to initiate a party or to court or serenade someone in public. The name derives from musicians' placement at a corner of a nearby street (*esquina*) to deliver their presentation. Rooted in Spanish serenades (*serenata*), the esquinazo became characteristic of the Creole auditory and social life in cities and towns during the nineteenth century. The sociocultural framework of Mariluán's esquinazo allows him to court his lover while simultaneously impressing the local White elite with his mastery of their sonic codes. In the setting of a "frontier city" such as Los Angeles, the esquinazo scene exhibits the seductive possibility of an assimilated Mapuche by appealing to the ears and eyes of the Creole society:

> Las personas del salón se agolparon a la ventana y a las puertas, mientras que las cuerdas de la guitarra principiaron a vibrar melodiosamente bajo los dedos de Mariluán, que cantó:
>
> Ecos del alma mía
> son mis suspiros;
> y para unirse a tu alma
> buscan camino.
> Tú eres la aurora
> y yo el valle que alumbra
> tu luz hermosa.
>
> Un aplauso unánime y estrepitoso estalló al apagarse las últimas vibraciones de la dulce voz con que Mariluán había entonado aquella estrofa. (Blest Gana [1862] 2005, 27–28)
>
> The people of the salon rushed to the window and the doors, while the guitar strings vibrated melodiously beneath his fingers and Mariluán began to sing:
>
> My sighs are
> Echoes of my soul
> Searching for the path

To unite with your soul.
You are the aurora
And I the valley illuminated
By your beautiful light.

Unanimous and raucous applause exploded with the final vibrations of the sweet voice with which Mariluán had sang that verse.

In this sonically festive segment, Mariluán's physical presence and musical virtuosity are praised through an outburst of vibrations. His esquinazo performance seems to rhythmically concatenate with the resounding of "the guitar strings," the "unanimous and raucous applause" from the audience, and the dulcet tones of his "sweet voice." The lyrical "echoes" and "sighs" of his verses strengthen this "melodiously" vibrant setting. Mariluán's performance of a song that, as the narrator remarks, he himself has written, confirms his competence in the art of the guitar and vocal interpretation. It shows his mastery of Western techniques of language and representation such as writing, the esquinazo musical genre, and a Europe-originated chordophone instrument, which by the mid-nineteenth century are components of the emerging Chilean Creole culture. Again, the role of vocal expression is key as the Indigenous character's "sweet voice" crowns Mariluán's performance before the White local elite.

Blest Gana's assimilationist fiction thus imagines the possibility of a Mapuche body able to replicate the vocal system of a Creole musical genre by following the aural standards of the settler Chilean script. In this nineteenth-century Creole setting, the Indigenous ability to vocally, socially, and ontologically act as a "human" was constantly under suspicion. Referring to the status of popular subjects in nineteenth-century Colombia, Ochoa Gautier states that "the voice needed to be regulated in order for that human to become a proper person" (2014, 203). In this case, a similar "acoustic biopolitics" (Ochoa Gautier 2014, 150) seems to be allegorized through Mariluán's musical performance. Blest Gana depicts an Indigenous persona who fits well in the aural script of a wingka society that favors a "sweet," complacent type of Native vocality that the colonial ear longs to hear and assimilate into its sonic, cultural, and social codes, especially as a resonant illustration of the so-called Pacification of the Araucanía.

Blest Gana's narrative shifts from the military and settler-family ambiance of a frontier city to Mariluán's community of origin. Yet Blest Gana had never traveled outside Los Angeles; he never experienced a Mapuche community. Instead he relied on hearsay, a repertoire of oral stories and conversations gathered during his visit to La Frontera, plus letters, documents, and books from his colonial library. His fictive entrance into the interior of La Frontera drew on that repertoire and on the fantasy and exoticism that permeated nineteenth-century Chilean and Latin American Creole literary representations of Indigenous territories. The rhetorical modes he used to present community authorities (the caciques) and the male warriors, mostly young people (*mocetones*), with whom Mariluán plans to meet, exemplify Blest Gana's reliance on his own imagination:

> Estos y aquellos formaban el número de cincuenta guerreros, armados de lanza y formados en semicírculo. Al centro esperaban los caciques, montados en hermosos caballos. Los resplandores de la luna iluminaban con tintes misteriosos aquellos rostros pálidos a los que las negras cabelleras flotantes hasta los hombros y sujetadas en la frente por cintillos rojos, en algunos, y en otros por cintillos de metal dorado, daban un aspecto imponente y fantástico. (Blest Gana [1862] 2005, 49)

> Fifty warriors armed with lances stood in a semicircle. At the center, the caciques waited, mounted on beautiful horses. The mysterious colors of the moon shined upon those pale faces, which the shoulder-length flowing black hair contained by gilded metal bands lent a magnificent and fantastic appearance.

Blest Gana lets the Mapuche appear as "warriors" whose well-organized layout in "semicircles" and "on beautiful horses" recalls Ercilla y Zúñiga's epic. Yet this description soon gives way to a pictorial rhetoric characteristic of Native settings in post-Independence novelistic conventions. Indeed, the passage is imbued with that genre/register of language recurrent in realist and naturalist literary portraits. Adjectives chisel the contours of nouns to create a pictorial effect: "mysterious colors," "pale faces," and the "magnificent and fantastic" aspect that imbues the image. Through this rhetoric of the exotic, Blest Gana's literary portrait of the Mapuche exceeds its realist, naturalist parameters to become romantic and gothic. By highlighting their "pale faces" and "flowing black hair" under the shining moon, the omniscient narrator casts a moribund and nocturnal spell around the bodies and shadows of the Native mass.

The paleness of the face recalls visual representations of the corpse typified in the European gothic imagination of the eighteenth century. It also recalls the trope of the "pale-faced Indian" that proliferated in the visual narratives of twentieth-century cinema throughout the Americas. The multilayered pictorial and literary aesthetics that run through Blest Gana's novel unveil the aura of mystery and death that framed representations of La Frontera's "interior" in the nineteenth-century Chilean Creole imagination.

As a connoisseur of the genre, the Chilean author also invests in a well-woven construction of the conflict in the novel's plot by adding the complexity of multiple perspectives and voices on La Frontera. Given Blest Gana's hybrid alignment with the realistic novel and his interest in situating the narrative at the intersection of fiction and history, the multiple perspectives invest the fictive discourse with the expected political discourse of the Mapuche against the Chilean colonial yoke. Halfway through the novel, the protagonist delivers a powerful speech before his Mapuche community. Put in quotation marks by the narrator, Mariluán's words read and "sound" as follows:

> "Tenemos derecho de conservar nuestro territorio y el sagrado deber de combatir por la defensa de nuestras familias. Os ofrezco mi vida para esto y pido solo el mando general durante la guerra. Quiero que la obediencia sea sin réplica, sin reflexión el arrojo, sin flaqueza la constancia. Si muero, mi hermano Cayo podrá continuar mi obra. El fin a que aspire llegar es el siguiente: que el Gobierno de Chile reglamente la internación de sus súbditos en el territorio de nuestros padres; que las autoridades nos presten su amparo, comprometiéndonos nosotros a respetarlas; que nuestros hermanos sean devueltos a sus hogares, y que se nombre tribunales que oigan los reclamos que tenéis que hacer contra los que os han despojado de vuestras tierras." (Blest Gana [1862] 2005, 49–50)

"We have the right to conserve our territory and the sacred duty to fight to defend our families. For this, I offer you my life and only ask to be the commander of the war. I want unflinching obedience, unsparing audacity, and unyielding commitment. If I die, my brother Cayo can continue my work. My aspiration is this: that the Government of Chile regulate its citizens' entrance into the territory of our parents; that the authorities offer us help in exchange for our pledge to respect them; that our brothers be returned to their homes; and, that hearings be called to hear your complaints against those who have robbed your lands."

This extraordinary passage leaves readers confident in Mariluán's commitment to the Mapuche people's struggle for territorial sovereignty in the mid-nineteenth century. Through the voice of his main Indigenous character, Blest Gana expresses unquestioned support for Mapuche collective rights to their "territory" and for their rebellion as a "sacred duty." Blest Gana's novel thus opens space for a Mapuche voice that conveys a sense of resistance and liberation that can be equated with the Indigenous politics of collective rights, land restitution, and the struggle for self-determination of the late twentieth and early twenty-first centuries. For a book of the early 1860s, this eloquently affirmative statement conveys a striking literary openness to Mapuche land politics.

The passage led Jorge Pinto Rodríguez to praise *Mariluán*'s historical context. "It is a literary work published in 1861 when the newspapers *El Mercurio* from Valparaíso and *El Ferrocarril* from Santiago were in full campaign to discredit the Mapuche" (2003, 261). Through this prism and his analysis of few segments of the novel, Pinto Rodríguez highlights how Blest Gana portrays a Mapuche novel's multiple voices in its early sections and thus allows the Mapuche perspective of land sovereignty and liberation to surface through Mariluán's speech: "We have the right to preserve our territory." Mariluán voices the "goal" of their struggle to his Mapuche peers, calling on them to demand concrete responses from the Chilean settler authorities, including due process to clarify the ongoing colonial practices of land dispossession in Ngulu Mapu: "to hear your complaints against those who have stolen our lands." Is this voicing of the Native demands enough to characterize the overall perspective of the novel? Although I agree with Pinto

Rodríguez on the progressive stance of the cited passage, this partial segment in no way establishes an overall positive view of Blest Gana's approach to the Mapuche. A novel constitutes an integrated whole, to be analyzed from beginning to end as a totality of relations. My analysis examines the prevailing role of a colonial ear that frames Mapuche sounds, characters, and lifeways in Blest Gana's narrative saga. The acoustic and literary disfigurement of the Mapuche grows progressively dominant as we approach the middle and end of *Mariluán*.

First, though, the matter of language itself surfaces in Mariluán's speech. The quotation marks enclosing the Indigenous protagonist's speech suggest a sort of transcription by the narrator, to be fully textualized in Spanish. This detail is striking, for in earlier paragraphs Blest Gana's omniscient narrator notes that the discourse between the protagonist and his people was delivered "with the rhythmic and severe accent of his native language" ([1862] 2005, 49). The character's "harangue" has been expressed "in the same language" as the "caciques." Although Mapudungun is the linguistic code of the Indigenous characters in the novel, the Mapuche language is nowhere in the book; Spanish is fictively naturalized as the language of Ngulu Mapu. In form and communication code, and without any apparent discontinuity, monolingualism dominates the apparent dialogic space of *Mariluán*'s narration. Spanish takes over the novel on a Mapuche subject, and Mapudungun remains nameless, evicted from the lettered realm. The "native language" exists only on a referential level. Throughout the novel, the omniscient narrator's ear and pen are open only to the wingka language; he erects a linguistic wall that affirms his position on the colonial side of La Frontera.

THE EAR AS A COLONIAL BORDER

Blest Gana's linguistically assimilationist script evolves in parallel with the increasing conflict between the Mapuche protagonist and the two major agents of settler colonial expansion that prevail in the novel, namely, the Tudelas—a Creole family—and the Chilean army. The deterioration of the omniscient narrator's empathy for the Mapuche becomes more evident. By the middle of the novel, the Mapuche-Chilean romance between Mariluán and Rosa reaches crisis, as does the Mapuche society–Creole colonial society relationship. Rosa's older brother, Mariano Tudela, opposes his sister's romantic relationship, stating, "We have declared war against each other" (Blest Gana [1862] 2005, 4). As the relations between Chilean settlers and the Mapuche communities near Los Angeles sour, Mariluán voices his decision to return to his native lands to organize a Mapuche uprising. His words lead him to a course of action, and the plot veers into open conflict. The military confrontation between the Chilean army and Mariluán's community forces the narrator's colonial ear to reveal its own representational limits:

> Al ruido de las armas se unían el chivateo general de los indios, el movimiento de los caballos y las voces de los jefes, aumentándose la confusión con el empeño de los oficiales en organizar los pelotones para atacar a su vez. Mariluán conoció entonces que prolongándose la resistencia y estableciéndose el orden entre los contrarios, perdía, en caso de ser derrotado, la oportunidad de retirarse, y dio la voz a Cleu, que tocó la señal convenida. (Blest Gana [1862] 2005, 52)
>
> The Indians' *chivateo* joined the noise of the weapons, the horses, and the voices of their chiefs, all of which increased the Chilean officials' confusion as they struggled to organize their troops in order to attack. Mariluán realized then that if resistance continued and his enemies were able to establish order, he would lose any hope of retreat if he were defeated. So, he gave the order to Cleu, who sounded the agreed-upon signal.

Here Blest Gana's narrative persona stigmatizes the Mapuche collective way of cheering and encouraging each other before battle by using the term *chivateo*. A Chilean idiomatic expression, this noun derives from the verb *chivatear*, which means "to shout, to vociferate" (Real Academia de la Lengua Española 2001). Yet a much more semantically accurate translation would be "screaming and vociferating like goats," since *chivateo* etymologically comes from *chivato*, that is, *chivo* or *cabrío* (goat). By equating a Native form of cheering with the bleating of goats, Blest Gana's narration, from its own anthropocentric angle, derogatorily assigns the Mapuche the dehumanized and primary mode of expression of a "primitive" being.

Blest Gana is not alone in this stigmatization of Mapuche modes of interjection. In Chilean lettered discourse, this practice of listening and representing recurs from the nineteenth century onward. The lesser-known Creole writer and military official Alberto del Solar also uses the epithet *chivateo* to describe the cheers of the Mapuche in his novel *Huincahual: Narración araucana*, published in 1888. Although written and published in the late nineteenth century, del Solar's novel adopts the historic period of Spanish colonialism as its fictive scenario. Set in and around "Villa-Rica," an emblematic settler town in Mapuche territory, the novel opens with the scene of an assault over this village (*malón*) by an army of "indios" led by the protagonist, Huincahual, who ends up kidnapping a Creole woman.[19] According to the narrator, the woman was "laid like a bundle across the saddle of her savage captor, the terrible Huincahual, son of the *toqui* Paillamachú, caudillo and lord of a hundred legions of Huilliches" (Del Solar 1888, 12). Earlier, the narrator had depicted the arrival in the village of the "ferocious Indians" and their "impetuous" shouts:

> El mismo clamoreo, ya muy cercano, y el eco de los cascos de un tropel de caballos desbocados se mezclan á trechos con el grito horrible, destemplado, del salvaje *chivateo* de los jinetes, ansiosos de botín, rabiosos de venganza, llegando hasta el corazón mismo del pueblo y llenándolo de terror y espanto. (Del Solar 1888, 8; italics in original)

> Now much closer, the raucousness and the echo of the hooves of a herd of runaway horses mix sporadically with the horrible, shrill scream of the savage *chivateo* from the horsemen eager for booty, thirsty for vengeance, who approach the very heart of the town, filling it with terror and fear.

Like Blest Gana, del Solar stigmatizes the vocal expressions of the Mapuche with the trope of the *chivateo*. References and adjectives that barbarize the Indigenous subjects and their surroundings strengthen its derogatory use: "runaway horses," "the horrible, shrill scream," horsemen "thirsty for vengeance," "terror and fear." Throughout *Huincahual*, such terms are always attached to the "ferocious Indian." In contrast, in the opening lines of his narrative, del Solar characterizes members of the Hispanic colonial settlement of Villa-Rica as "noble and genteel" (1888, 1). Through the supposedly "impersonal" and soft voices of their omniscient narrators, Blest Gana and del Solar's narratives instantiate not only the colonial edge of their aurality but also what scholar Jennifer Lynn Stoever characterizes as "the unspoken power of racialized listening" (2016, 7).

This colonial and racial propensity to hear Mapuche expressions as *chivateo* permeates literary works of Creole writers of the second half of the nineteenth century and appears in several dictionaries devoted to Chilean slang. In the classical dictionary of *Chilenismos* (Chilean expressions) (1928), scholar José Toribio Medina glosses the term *chivateo* as "vocinglería, gritería desaforada y grosera" (shouting, riotous and vulgar screaming). In subsequent lines, Toribio Medina's definition of the verb *chivatear* associates it explicitly with the Mapuche: "Dar gritos desaforados al tiempo de acometer un bando o ejército enemigo al otro, práctica araucana" (Uncontrolled screaming when attacking an enemy side or army by another; Araucanian practice) (1928, 119). In his "Chilenismos de uso corriente" (Chilean expressions of common use) (1965), Naranjo Villegas emphasizes the nonhuman animal dimension of the term *chivateo*, defining it as "bullicio producido por muchos individuos. Por analogía con el sonido que emiten las manadas de chivos" (din produced by several individuals. By analogy, associated with the sound emitted by a herd of goats) (1965, 609).

Clearly, Blest Gana's trope of chivateo resonated with Chilean settlers' idea of the Mapuche people as barbarians or savages. Further, Blest Gana's stigmatization bolsters a Chilean ideology of colonization that requires the presence of the "savage," or the *auka*, a Mapudungun term used in that sense in Spanish colonial chronicles. The notion that Mapuche society lacked order and governance underpinned the justification of the so-called Campaign of Pacification of the Araucanía, as well as the subsequent politics of dispossession and state, oligarchic, and ecclesiastical tutelage over Native communities.

The passage on chivateo from Blest Gana's novel also makes clear how a marked Chilean colonial ignorance about the sonic expressive acts of the Mapuche underlies their stigmatization and disfigurement. The passage closes with Mariluán's order to Cleu, one of his men, to announce the withdrawal of the Mapuche warriors from the battlefield through what the narrator refers to as an "agreed-upon signal" (*señal convenida*). The narration then attempts to specify: "El sonido de la corneta hizo ponerse a los indios en marcha, acostumbrados a esta clase de escaramuzas" (The sound of the bugle set the Indians in motion, accustomed as they were to this kind of skirmish). This raised among the men of "the Government" (of Chile) the presumption that "that sound ordered some new maneuver" (ese sonido ordenaba una nueva maniobra) (Blest Gana [1862] 2005, 52). For Blest Gana, "that sound" evokes a Western-style bugle (*corneta*). But it more likely came from a Native wind instrument such as the *kull kull*, commonly used to alert people to gather or get moving.[20] To Blest Gana's Creole characters, as well as to the narrator himself, this sound and other acoustic and verbal expressions of the Mapuche are inaudible, or they are heard through a filter that lacks specificity.

Limited knowledge of the Mapuche people underlies Blest Gana's narration. While this is paradoxical for a writer aware of the principles and conventions of literary realism, it is symptomatic of a larger issue of colonial ignorance in nineteenth-century Chilean society. This ignorance leads his novel to resort to stigmas, to racialize Native sounds, and to represent the Mapuche as an infrahuman mass all too accustomed to "skirmishes." In contrast, the Chilean soldiers and army associated with the "Government" represent law and order. Through this Chilean colonial prism, the male warriors led by Mariluán belong to an Indigenous society prone to needless scuffles, lawlessness, and violence—a Wallmapu immersed in backwardness and barbarianism.

Mariluán thus ultimately confirms the stigma of the Mapuche as a "primitive race" (Blest Gana [1862] 2005, 117). As the plot develops a series of actions in which the Native protagonist is torn between the struggle of his people and his romance with Rosa Tudela, he faces puzzling borders between Mapuche society and Chilean society, politics and love, reason and emotion. In these fictive crossroads, Mariluán and other Mapuche characters succumb to their primary instincts. Thus, it is a fellow Mapuche, Peuquilén, who ends up betraying and murdering Mariluán because of his conflicting, uncontrolled desire for the same White Chilean woman (Rosa). Peuquilén's assassination of Mariluán closes the novel.

> Durante esta atroz operación, los ojos de Peuquilén brillaban con los sombríos resplandores de la venganza satisfecha. El temor que le inspiraba Mariluán había desaparecido. A la escasa luz de las estrellas contempló el rostro pálido de su víctima y sus facciones se iluminaron con una salvaje alegría; para él, la cabeza de Mariluán representaba la satisfacción del rencor y el pago ofrecido al asesino por las autoridades chilenas. La helada sombra del remordimiento no oscureció por un solo instante la expresión de salvaje crueldad con que Peuquilén sostenía de los cabellos la sangrienta cabeza de Mariluán. (Blest Gana [1862] 2005, 117)

> During this terrible operation, Peuquilén's eyes glowed with the somber shine of satiated vengeance. His fear of Mariluán had disappeared. Under the faint light of the stars, he contemplated the pale face of his victim, and his features illuminated a savage happiness; for him, Mariluán's head represented the satisfaction of rancor and the payment offered to the assassin by the Chilean authorities. The cold shadow of regret did not for an instant darken the expression of savage cruelty with which Peuquilén held the hair of Mariluán's bloody head.

The novel leaves us with an image of an Indigenous society unable to regulate its "savage" tendencies, iteratively embodied through and around the figure of Peuquilén. Either repeated as "savage happiness" or as "savage cruelty," savagery is endemic to the Mapuche in Chilean discourse.[21] Blest Gana echoes a vox populi in the Creole mainstream: "The Indians kill themselves!" The anti-Indigenous stigma of barbarism, continuously insinuated throughout the Chilean author's narration, suffuses the paragraph above. Even more, Blest Gana's narration highlights that, prior to his last breath, the agonizing Mariluán emits "un rugido de despecho" (a spiteful roar) in the middle of the dark native forest. Thus, the barbarized acoustic of the "roar" hurls Mariluán into a liminal expressive space between human and nonhuman animal. In that nonverbal liminality, he is figured and disfigured as a being similar to those species of the Ngulu Mapu forests that sound equally "savage" to the settler colonial ear. As soon as Mariluán's acoustically barbarized agony ended, the panorama turned even more terrifying, as "everything fell into the most profound and lugubrious silence" (todo quedó en el más lúgubre y profundo silencio) (Blest Gana [1862] 2005, 116).

By highlighting the "pale face" of Mariluán's corpse, a visual image that the narrator also uses to describe the Indigenous warriors' countenance, Blest Gana's narration traces a contiguity between paleness and the Native body, placing the latter in a ghostly and lifeless realm. This dehumanizing image of indigeneity in settlers' imagination connects to the later pervasive visual and corporeal cinematic trope of the "pale-faced Indian." Within this death imaginary, Blest Gana's novel concludes by coupling indigeneity and silence. The novel ends in a scene of elimination and silencing: the cut throat and the headless body of the Mapuche protagonist, a Native leader for whom life as voice has been canceled, as one of the concluding phrases makes clear: "Ni un solo grito, ni una sola exclamación fue lanzada por el infeliz caudillo de los araucanos" (Not even one scream, not even an exclamation came from the disgraced chief of the Araucanians) (Blest Gana [1862] 2005, 116).

Reduced to a state of nonverbal expression, the Indigenous body is divested of articulated voice and speech. This scene of agony and death allegorizes the reality of a silenced Mapuche people, headless and voiceless, emphasizing the destructive and barbarian forces that allegedly underlie Native lifeways. What Ochoa Gautier calls the "spectrality of sound" (2014, 8) here becomes that of silence. Subsequently, as literary scholar Alvaro Kaempfer describes it, in this narrative "the final erasure of the native subject allows its entry into history only as a subaltern and spectral figure" (2006, 89).

Such racialized fantasies confirmed for the Chilean reading elites of the nineteenth century their view of La Frontera and the Mapuche people as a world of "endemic" destruction and death. Hence, by default, they had no reason to question the "Pacification Campaign." In Blest Gana's novel, wingka male settlers, as novelistic characters, remain safe from the bloody and turbulent events of Ngulu Mapu. In contrast, the fictive closure of *Mariluán* leaves readers with a lugubrious image of Mapuche life. By the end of the novel, all emphasis is on the dark instincts of "rancor" and "savage cruelty," treason and violence, that allegedly destroy the fate of Native characters. Politically, too, the final image of Mariluán's corpse reinforces an ideology that sees the Mapuche as "headless" and submerged in chaos. The headlessness marks the elimination of the possibility of speech and language from the Indigenous body, the body of a Mapuche who had achieved the status of "head" (longko) of his own community.

A parallel to the silent, pale body of the Mapuche protagonist is the transformation of the White Chilean Rosa Tudela, traumatized by the fate of her Mapuche lover and crushed by the gendered, racial, and social powers of the Creole society epitomized by her brother. In a fatherless home, Mariano imposes a masculine domination that controls Rosa's affairs as well as the Tudela family's business, while isolating her and the family from Mariluán's amorous and political affairs. In this melodrama of colonial and patriarchal power relations, neither the Indian nor the woman achieves a happy ending. Blest Gana ends his novel by citing a letter from Juan Valero, a non-Indigenous male character who seems to have been a loyal friend of Mariluán. Valero writes to a friend from Santiago about the tragic events in Los Angeles. Indeed, the closing paragraphs of the novel consist of a full transcription, put in quotation marks, of Valero's letter; thus, he becomes the first-person narrator who informs us of the final fate of Mariluán as well as that of Rosa Tudela. Valero's portrait of the female character suggests that she ends her days as a living corpse. In the letter's last paragraph, which becomes the closing lines of the novel, Valero writes,

> "Todos los días voy a informarme de la salud de Rosa. Al tercero cesó el delirio y desde entonces vive sentada en una silla, sin mirar a nadie, sin hablar una sola palabra y dirigiendo de cuando en cuando la vista hacia la calle. A la hora en que vio la cabeza de Mariluán, un gran estremecimiento sacude su cuerpo, da un grito agudo y permanece delirando dos o tres horas; después cae en el abatimiento profundo de antes. En estos diez días transcurridos desde la hora fatal en que salimos juntos a la puerta de calle, su persona ha sufrido una completa transformación; parece un cadáver y es muy de temerse que no sobreviva mucho tiempo al peso del dolor." (Blest Gana [1862] 2005, 120–21)

> "Every day I go to Rosa's home to check on her health. By the third day her delirium ceased and ever since, she passes her time sitting in a chair, without looking at anyone, unable to utter a single word, turning her gaze now and again toward the street. When she saw Mariluán's head, her body writhed. She let out a sharp scream and fell into a delirium for two or three hours before returning to her deep gloom. In the ten days since the fatal moment when we went out the door to the street, she has undergone a total transformation; she resembles a corpse. It is feared she will not survive the burden of her pain for long."

If the headless body of Mariluán symbolizes the Native voice's end and marks a closure in the novel, the above passage underscores a similar condition for the female character. Rosa becomes a spectral figure, split between prolonged silences—"unable to utter a single word"—and the "scream" of her own delirium. Like the Native subject, Rosa "resembles a corpse," with no prospect for survival, condemned to the nineteenth-century fate of a femme fatale. Both the Indian and the woman are tossed over a temporal precipice. "Savage" and "woman" represent a racial and gender anomaly in the social, economic, and military order of the Creole masculine, colonial, and capitalist dominion emerging in La Frontera.[22] It is inevitable, therefore, that they share the fate of voicelessness, as the wretched people of history, and are relegated to death, literally and allegorically.

LITERARY PACIFICATION?

Given the context of the 1860s, Blest Gana's novel *Mariluán* literally and allegorically operated as another mechanism of war in the Chilean settler-state campaign against the Mapuche people in this period. It perpetrated discursive and representational violence on Ngulu Mapu, ranging from its initial assimilationist narrative to the prevailing language of acoustic disfigurement of the Mapuche as savages. Recall that Blest Gana studied literature and humanities but also underwent years training at the Military School of Santiago. This background no doubt influenced Blest Gana's approach to the Mapuche people. Based on this lettered and military formation, he traveled from Santiago to Los Angeles. First, as a writer influenced by literary realism, Blest Gana aimed to fulfill one of the key principles of this aesthetics: the practice of observation. Second, his novel offers subtle depictions of how the Chilean military acts under a regime of discipline, order, and honor. In contrast, chaos, treason, and blood characterize Mapuche society. In a 1955 review, influential literary critic Raúl Silva Castro highlights how *Mariluán* "succumbs to the desire to accumulate tragic and bloody effects to provoke impressions of horror in the reader" (1955, 205). Silva Castro criticizes Blest Gana's melodramatic tendencies within the aesthetic domain, but without delving into the ideological layers of the novel's "bloody" sensationalism in its representations of the Mapuche. In my view, this sensationalism is not merely a literary or stylistic inclination; rather, it is ideologically strategic. Through the elevation of a "tragic and bloody" rhetoric in his prose, Blest Gana's novel ultimately stigmatizes Ngulu Mapu as a land of endemic violence, an idea that even today, in the early twenty-first century, prevails in the Chilean settler imagination. Furthermore, Blest Gana's anti-Indigenous racial ideology and belief in White Creole supremacy become explicit through the narrator of a previous novel, *La aritmética en el amor* (1860). Making reference to the Chilean Constitution of 1828, he writes,

> La Constitución abolió los títulos, mas no pudo abolir la nobleza por dicha nuestra, sin la cual nos veríamos en la dura precisión, de no encontrar un solo *caballero* a quien dar la mano por esas calles de Dios. Y bien que muchos pretendan que no es la ilustración y brillo intelectual lo que esas familias nobles se han encargado de perpetuar, puede a los tales respondérseles que en cambio han conservado la pureza de la raza, lo que es una base de progreso en todo país sensato, y van transmitiendo a sus herederos la blancura del cútis, sin la cual cualquiera podría tomarnos por verdaderos indios, sin que nos quedase el derecho de ofendernos por tan insultante equivocación. (Blest Gana 1860, 85–86)
>
> The Constitution abolished aristocratic titles, but fortunately it was unable to abolish nobility, without which we would be in the harsh circumstance of not finding any *gentleman* to shake hands with on those God's streets. Although many allege that these noble families perpetuate neither Enlightenment nor intellectual brilliance, they have preserved the purity of the race, the foundation of progress in any sensible nation, and they pass on to their heirs the whiteness of their skin, without which we could be taken for real Indians, unable to defend ourselves from such an insulting mistake.

Even in the mindset of a nineteenth-century lettered liberal like Blest Gana, the imagining of the Chilean nation is clearly tied to the supremacy of *la blancura* (whiteness). *Mariluán* forms part of the same historical process that aimed to consolidate a settler colonial nation-state under White Creole dominance. When Blest Gana published *Mariluán* in 1862, Chilean military chiefs had already presented the government with several plans for occupying and controlling the southern territories still under Mapuche governance. Colonel Cornelio Saavedra submitted his own strategic plan in early 1861, as did General Cruz in April of the same year. As a corollary to these governmental deliberations, on December 1, 1862, Chilean military forces under Saavedra's command arrived in the vicinity of the frontier town Angol to establish possession and rebuild a colonial urban settlement that Mapuche forces had repeatedly dismantled during the seventeenth century (Pinto Rodríguez 2003, 237–38). In the very year Blest Gana's novel was published, the so-called Campaign of Pacification of the Araucanía began to take shape.

The temporal and geographical correspondence of Blest Gana's novel with the Chilean invasion of Ngulu Mapu entangles literature and history in the same plot. *Mariluán* ends with a scene of death: the silencing of the Indigenous protagonist allegorically stages the colonial occupation of Mapuche lands and communities by military, economic, political, and ecclesiastical agents of the Chilean nation-state between 1862 and 1883. Blest Gana's plot both parallels and prefigures the so-called Pacification Campaign, a concerted Chilean state action to subjugate and silence the Mapuche. Tellingly written in the late 1850s and early 1860s, it both registers and connotes closure. With the beheading of the Mapuche protagonist "under the scarce light of the stars" (Blest Gana [1862] 2005, 117), the book's final pages become a tombstone. As part of the Chilean colonial library, *Mariluán* performs the beginning of a *literary pacification* over the voices of the Mapuche.

Throughout Blest Gana's novel, the colonial ear constitutes an omnipresent force, be it to figure or disfigure, to include or exclude the Mapuche under its own terms. In this vein, the novel reveals how literature became part of the Chilean colonial settlement by establishing a way to hear, codify, and mediate Mapuche voices and sounds. Literature, drawing, painting, and photography played significant roles in the articulation of aural and acoustic modes of colonial mediation, assimilation, and erasure in the second half of nineteenth century. As the next chapter discusses, new media of sound reproduction, such as radio and the music recording industry, replicated the deployments of acoustic colonialism within the mediascape of the twentieth century in Chile, and over Ngulu Mapu.

TWO

Indio Pije: A "Mapuche" in the Mediascape

Sudden interruption: Chile, September 11, 1973. At the time, I was still a child under the care of my grandmother and my two uncles in Tralcao. In the early morning that day, we heard the news on the radio: The military junta headed by General Augusto Pinochet had assumed power, promising to restore "peace" and "order" to justify its overthrow of socialist leader Salvador Allende and his supporting left-wing coalition, the Popular Unity. Repression, censorship, and human rights violations started to dominate the country under the rule of a civic-military dictatorship that lasted seventeen years. Many radio stations associated with Allende's government were dismantled. Among them, Radio Corporación, the station that, during the 1960s and early 1970s, had been broadcasting Residencial La Pichanga, *a comic radio theater that we often listened to.*

After 1973, the mediascape changed dramatically and Radio Colo Colo, a Santiago-based AM *station that transmitted as* CB *76, had gained nationwide popularity. In those years, every night we tuned in Radio Colo Colo. In a nod to the "popular," the station adopted the name of Chile's most influential professional soccer team, Colo Colo. Furthermore, this is a name with a longer history in Ngulu Mapu: Colo Colo was a prominent Mapuche leader who fought against the Spaniards during the 1550s. With all these resonances, Radio Colo Colo became popular due to its festive musical programming, which included cumbias, Mexican rancheras, and Chilean cuecas. These danceable musical genres were extremely popular in Chile, not only among Creole and mestizo campesinos but also among the Mapuche in rural areas—especially among the Mapuche-Williche population, my Huechante Lefno family included. Radio Colo Colo was easy for most listener to tune in at night. Thus, "la Colo Colo" gained a vast audience in working-class urban barrios and rural areas.*

In 1975, Chilean actor, comedian, and singer Ernesto Ruiz debuted as Indio Pije with an LP *titled* Qué pasa en la ruca: Show de cumbias, *issued by the small record label Via Music. The titular song of the album, "Qué pasa en la ruca," became a hit in the programming of Radio Colo Colo. Thus, from those politically obscure years, I still remember the voice of Indio Pije festively resonating on our ears at night. The song title already had a Mapuche marker,* ruca, *which is a term from Mapudungun that means "home" or "house."*[1] *With lyrics depicting an "Indian" home environment of constant celebration and merriment, the cumbia song "Qué pasa en la ruca" helped popularize the idea of an endlessly joyous night fiesta of the Mapuche, paradoxically during the most repressive years of Pinochet's dictatorship.*

....

Several decades before the debut of *Residencial La Pichanga*, radio had played a significant role in the consolidation and expansion of Creole national projects and mass culture in Latin American societies (Merayo Pérez 2007, 11–22). Together with the populist politics of the 1930s and 1950s, radio played an influential role in the expansion of hegemonic "national cultures." It aurally connected popular masses from urban and rural areas. Media scholar Jesús Martín-Barbero notes that "radio became the domain of the popular, the realm of the oral" ([1987] 1993, 169). Across Latin America, radio deeply influenced the articulation of mass culture and a popular sense of "national" belonging. Radio "gave people of different regions and provinces their first taste of nation" (Martín-Barbero [1987] 1993, 164).

During the nineteenth century, written literature in books and newspapers was made possible by the technology of representation for settler colonial Latin American societies, especially for the Creole lettered elites. Certainly, written literature was a significant component of the "colonial library" (Mudimbe 1994) that accompanied the Chilean Creole elites in the process of settler nation-state building. Newspapers progressively gained mass readership throughout the twentieth century. However, although books and newspapers were available to the Chilean Creole upper and middle classes, they did not figure in the daily life of most of the Mapuche people, nor of the vast popular sectors of the settler Chilean society, in this period.

The emergence of radio, as it grew into a key component of mass culture in Chile by the 1930s and 1940s, significantly altered the mediascape of the country. Broadly speaking, as historian Christine Ehrick notes, "Radio was the Internet of its day." It revolutionized how "people received information and perceived their place in the larger human community" (Ehrick 2015, 2). Important studies have demonstrated that radio truly became a mass cultural force throughout Chile in the early 1960s by which time "a good number of radio stations cover[ed] almost the whole country, with the exception of Chiloé, Arauco and Aysén" (Lasagni et al. 1985, 14). Undoubtedly, by the end of this decade, radio had become an influential factor in the aurality and sonic environment of Mapuche homes and community settings in Ngulu Mapu as well.

The first radio transmission in Chile took place in the Universidad de Chile campus on August 19, 1922, while some two hundred people gathered around a receiver in *El Mercurio* newspaper's headquarters in downtown Santiago.[2] This pioneering broadcast reached other spots in the capital and in Valparaíso, where "wireless stations" had been put into operation. In March 1923, the country's first commercial radio station began broadcasting under the symbolic name of Radio Chilena (Lasagni et al. 1985, 5–6; Pastene 2007, 113–16). Transmitting from downtown Santiago, this station helped radio become an influential media tool for the *Chileanization* that the nation-state was aiming to consolidate. Over the following decades, this settler nationalism found an effective tool in radio for its propagation and impact, not only onto the Mapuche but also onto other Native territories.[3] This assertion of *chilenidad* erased indigeneity from the mediascape, establishing a narrowly crafted regime of public audibility and sonority.[4] Indeed, settler impersonation of Mapuche voices and bodies became an intrinsic part of the hegemonic mediascape, especially through music and *radioteatro* (radio drama, or radio theater).

Indio Pije, a supposedly Mapuche character featured on *Residencial La Pichanga*, a radioteatro de humor (comic radio show, or comedy radio theater), exemplifies the way in which mediation and representation of Indigenous voices took form in Chile's radiophonic culture. As radio itself gained appeal and influence in the country's mass culture around the mid-1950s, the radio comedy genre, which had been around since the 1930s, grew in popularity (Pastene 2007; Martín-Barbero [1987] 1993; Gamboa 2000). The genre benefited from the new sentimental and sensorial culture that arose from the interrelationship between radio, populism, and popular culture, expanding on the influence it had had earlier in the radiophonic ether since the 1930s.[5] Residencial La Pichanga then further took advantage of this state of affairs during the 1950s and 1960s by aligning the program with the zeal for soccer that was taking over Chilean mass culture by the mid-twentieth century.[6]

Despite occasional gaps in transmission, Residencial La Pichanga managed a lengthy broadcast life from Santiago-based radio stations, extending from its beginnings on Radio Agricultura toward the mid- and late 1950s to its most successful period, from the early 1960s to 1973, on Radio Corporación. Initially, its Indio Pije character was an extension of the mascot of the Green Cross soccer team, which appeared in the 1950s and was consistently used until the 2000s.[7] However, by the mid-1960s and the early 1970s, the program was so established in the lives of its radio audience that its character Indio Pije began appearing in several other popular media in Chile.

In addition to a radiophonic character, Indio Pije had a presence in print culture and music. In 1965, César Enrique Rossel, the radio comedy's creator and producer, started publishing a *revista de historietas*, the rough equivalent of a comic book. This publication, which came out under the title *La Pichanga*, occasionally featured Indio Pije on its pages (Rossel 1965a). In 1975, through the small record label Via Music, Ernesto Ruiz, the actor who played the Indian character, issued an LP titled *Qué pasa en la ruca: Show de cumbias* (What's Happening in the Ruca: Cumbia Show). The main voice and persona in this series of cumbia songs was Indio Pije. The staging of the radiophonic character through print format and music undoubtedly increased his popularity.

As portrayed on the radio program *Residencial La Pichanga*, Indio Pije is a Mapuche who has supposedly moved up the Chilean social and economic ladder, becoming a *pije*, a person who ostentatiously looks and sounds like the Chilean upper-middle class, or attempts to. However, Indio Pije's theatrical, high-pitched voice and rustic, broken Spanish makes him a cartoonish figure. It is obvious that his name and identity—"Indio" and "Pije"—constitute an oxymoron. Ruiz's highly theatrical performance of the voice and speech of the Mapuche stages a crude caricature of it in radiophonic, musical, and graphic culture.

To what extent does the figure of Indio Pije exemplify a practice of colonial mediation and acoustic disfigurement over the Mapuche in the Chilean mediascape? To unveil the deployment of this colonizing practice, I conceive of the domain of voice and sound as a battleground where social, political, cultural, racial, and historical positions and relations of power played out in the sphere of sonority, whether in literal, actual, or imaginary terms. Within this framework, the caricatured staging of Indio Pije—as projected through the voice and physical mediation of the Chilean Creole performer—is more than mere mockery. The colonial supplanting of the voice and body of the Mapuche is intentional. Ruiz's word choice and vocal qualities form the critical essence of the performativity of the supposedly crude, rustic speech and exaggerated "manhood" that epitomizes the Indianness of his character. To achieve his voice projection as Indio Pije, Ruiz fully utilizes his throat, vocal cords, tongue, and mouth, as well as the musculature and gestures of his face. The personification of Indio Pije entails a corporeal appropriation that is essential to "performance" as a practice. In the words of scholar Diana Taylor, "The body is one more medium that transmits information and participates in the circulation of gestures and images" (2016, 60)—and, I would add, voices and sounds. The very body of the Chilean Creole actor (its vocal inflections, facial contortions, and mouth movements) produces an imaginary Indian that the settler radiophonic script and its subsequent media extensions attempt to personify acoustically and physically.

This way of figuring and disfiguring, representing and misrepresenting, the Indigenous subject's voice is analogous to "redfacing" in the history of cinematic representation of Native Americans, particularly during the first decades of the twentieth century, when White American actors were routinely cast as Indians in Hollywood Westerns.[8] As Comanche media scholar Dustin Tahmahkera states, "Representations of redface entail images of and discourses about Indigenous Peoples as enacted primarily by non-Native characters that play 'Indian'" (2008, 325). According to Seneca film scholar Michelle Raheja, in its more literal versions, the practice of redfacing allowed "impostors" to "pander to audiences' expectations and desires for a neatly packaged, accessible, and stereotypical form of Native American identity performance removed from any sense of social or community obligation or responsibility." Subsequently, the people involved in this "form of redfacing" become part of "a representational field" that "has already been defined by the dominant culture" (2010, 143). Indio Pije, imagined as an assimilated Mapuche, similarly embodied many of the same characteristics of the problematic script.

Whether as a response to or a projection of the identity anxieties of settler societies, the objectification of Indigenous peoples has a long history in both elite and popular culture. It encompasses a variety of discursive and sociocultural practices ranging from the impersonation of "Native" roles in elementary school children's games to the imitation of "Indians" as subjects of entertainment and the butt of jokes in comedic performances, to the use of Indigenous "mascots" by sport teams (as Indio Pije was for the Green Cross soccer team in Chile), and the practices of organized White settler groups that appropriate the "Native" experience.[9] In *Playing Indian* (1998), an influential study of practices of (mis)representation, Cherokee historian Philip J. Deloria scrutinizes how a "playing Indian" culture has functioned in White American society. Deloria delves into how a subset of Americans searched for Indianness either as "object hobbyists" who were intent on collecting "authentic" artifacts or, later, in a somewhat more elaborate trend, as "people hobbyists" (i.e., non-Indigenous Americans interested in "real Indian people"). These hobbyists were especially invested in learning more directly about certain American Indian traditions, such as dance and powwow culture, to put them into practice in authentically "Native" ways. Deloria contends that a sense of "racialized authenticity" projected through Indianness instigated and underlaid most of these practices. Such people "played Indian to address longings for meaning and identity that arose from the anxieties of their times" (1998, 137–50, 151). The comic staging of Indio Pije as a pseudo-Indigenous character invented by a Chilean scriptwriter resonates strongly as "playing Indian," particularly by replicating the logic of objectification of Indigenous subjects.[10] Indeed, joking and playing stereotypically around the figure of Indio Pije was a constant in Ruiz's various impersonations and representations across media.

Even though Indio Pije was on the radio throughout much of the second half of the twentieth century, and after years of searching, I have never found a recording of a *Residencial La Pichanga* episode that includes the Indian's participation. This speaks more of Chile's precarious and limited radiophonic archive than of the program's widespread popularity. At the Audiovisual Laboratory of the National Library in Santiago on January 3, 2023, I was able to listen to a digitized copy of the radio comedy show as broadcast through Radio Minería in 1987. Thanks to the archive of a former radio controller at Radio Monumental in the early 1990s, I heard a sample of what the station played on the air. Neither of these audio tapes (one of 17 minutes, 42 seconds, and one of 7 minutes, 42 seconds) included Indio Pije's voice. But the LP of Indio Pije singing, which has survived, provides key archival evidence of the character in song.

Given the absence of radiophonic archives to document Indio Pije on *Residencial La Pichanga*, my own memory as radio listener fills in this gap. Additionally, my Mapuche method of allkütun allows me to attentively navigate these radiophonic memories, listen to the music, and even approach print materials as spaces of aural experience. The performances of Indio Pije reflected how Ruiz's ear aurally imagined a Mapuche voice. In the staging of his character, the colonial ear filters what sounds Indigenous, and what then gets projected on the airwaves. The staging of Indio Pije is one more sample of acoustic colonialism. Practicing allkütun around his figure, however, gives me the opportunity to unsettle how the settler listens and projects indigeneity in the realm of vocality.

FROM SOCCER TEAM MASCOT TO THE RADIOTEATRO AIRWAVES

Prior to coming alive in sound, image, and movement, Indio Pije had been the official mascot of the Green Cross soccer team since the 1950s. In 1965, the same year that the radio-theater program *Residencial La Pichanga* first went on the airwaves of Radio Corporación, the Green Cross soccer team moved from Santiago to Temuco and became Green Cross–Temuco.[11] There, in the urban core of historic

Mapuche territory, Green Cross–Temuco unashamedly displayed the mascot with the trarilonko (headband) and the *kultrun* (musical "instrument"), which signaled the character's "Mapuche" identification (see figure 2.1).[12] The creation of an "Indigenous" mascot for a Chilean team with an English name is itself a cultural paradox. Yet this out-of-context use of an Indigenous subject captured the racial history of subordination of the Mapuche people that had been omnipresent in public culture since the state-sponsored occupation of Ngulu Mapu in the late nineteenth and early twentieth centuries.

The Chilean mainstream had long oscillated between a desire to integrate Indigenous peoples, an assimilationist view instituted by both right- and left-wing governments, and a cultural and racial logic of contempt toward "Indios." These antagonistic ideologies underlay the equally contradictory figure of Indio Pije, with its amalgam of friendliness and mockery, which explains why graphic representations depict him in a hyperbolic manner: Note the magnified eyes, big nose, and open-jaw guffaw in figure 2.1. Regardless of his creators' intent, the visual rhetoric of exaggeration affirms the mascot as a buffoon. There is a similar logic in the use of Indigenous figures by sports teams across continents. In the United States, where American Indians have long fought this, anthropologist Pauline Strong states that "the continued use of pseudo-Indian sports symbols exemplifies and reproduces the subordinate place of Native Americans within the dominant society" (84). In the Anglo-Chilean imaginary that marked the identity of the Green Cross soccer team during its existence, the conversion of an Indian subject into a mascot—that is, "an object" to manipulate—aside from a cultural contradiction, constituted a grotesque form of racialization.

In this character's very name, the epithet *Indio* resounds with the derogatory bias that exists in Chile. At the same time, that racial stigma is rhetorically balanced by its coupling with the adjective *pije*, which standard Spanish dictionaries such as the Real Academia de la Lengua Española define as "a man who dresses with excessive elegance and refinement" (Real Academia de la Lengua Española 2001, 1758). In Chile, the term itself can be traced to mid-nineteenth-century popular slang. Scholar José Toribio Medina, in his classical dictionary *Chilenismos: Apuntes lexicográficos* (1928), quoting a text from the period, highlights the figure's snobbishness but also states that a pije is a man who "somewhat aggressively looks at women whom he tirelessly chases" (288). These definitions capture the gender and sociocultural dimension of the epithet; yet neither the Spanish

FIGURE 2.1 · Indio Pije in action, whooping, as mascot of the Green Cross soccer team (renamed Deportes Temuco in 1986). Deportes Temuco team's website.

dictionary nor Toribio Medina captures the term's class bias, especially as used since the 1960s. In the early twentieth century, Manuel Antonio Román's *Diccionario de chilenismos y de otras voces y locuciones viciosas* addresses the class dimension. For Román, a pije's main characteristic is presumption, regardless of his socioeconomic background (1913–16, 276–78). Again, in the late twentieth century, glossaries of Chilenismos tend to bifurcate between the cultural and the socioeconomic. The Academia Chilena de la Lengua's *Diccionario del habla chilena* (Dictionary of Chilean Speech) characterizes a pije as a "person who presumes to be refined and elegant without actually being so" (Academia Chilena de la Lengua 1978, 173). Likewise, in their *Diccionario ejemplificado de chilenismos* (1987) Chilean scholars Félix Morales Pettorino and Oscar Quiroz Mejías offer a more updated definition of pije: "member, follower, or player of the Club Deportivo Green Cross (which is headed by people from the society's upper class)" (1987, 3609). Recurrent in these definitions is that pije, as a social prototype, emerged from the milieu of the settler Creole society in Chile. As a character born in the second half of the twentieth century, even though he seems to be more tied to the two latter definitions, the Indio Pije carried all the layers of meaning and disparate semantics of the term across time.

In conflating the racially derogatory epithet *Indio* with the white-washing term *pije*, associated with the Green Cross soccer team, Indio Pije oxymoronically embodies the idea of a Mapuche assimilated into settler Chilean capitalist society. The team's English name already signals the pro-Anglo, Europeanizing ideology and language of sports culture in Chile in which Indio Pije is culturally and racially subsumed. Thus, with all the allusions of his oxymoronic identity, Indio Pije became part of the cast of characters on the radioteatro *Residencial La Pichanga*. He fit right in, as the radio show was largely comprised of comical skits about the Chilean soccer league and featured several male Creole actors personifying representatives of professional soccer teams from Santiago and the provinces.

Between the 1960s and 1970s, through *Residencial La Pichanga*, Indio Pije's radiophonic voice gained a significant resonance in the mediascape. Radio dramas got more airtime on Santiago radio stations with nation-wide reach, such as Agricultura, Minería, Portales, and Corporación. Radio listeners throughout the country could tune in to these stations on AM at night and on shortwave during the day. Many provincial radio stations connected to or retransmitted the bigger urban stations' news programs and radio dramas. The increased popularity of radio dramas during the 1960s led to more programming time on several radio stations. Radio historian and broadcaster Luis Gamboa recalls that in that period, Radio Agricultura "managed to have nine radio dramas" a day (interview by author, 2024). In this context, *Residencial La Pichanga* directly competed for audiences against comedy radio shows with broader popularity in the Chilean mediascape. Among them, Gamboa mentions comic programs such as *Hogar dulce Hogar* and *La bandita de Firulete*, broadcast by Radio Portales, and *Radiotanda* on Radio Minería.

The very subject of soccer—the most popular sport in the country—helped *Residencial La Pichanga* establish its own niche in a highly competitive environment. The characters of this radio comedy lived in Santiago in a boardinghouse (*residencial*) while they participated in meetings of the National Association of Professional Soccer as representatives of the different teams. The show's dialogue mostly consisted of masculine characters making jokes at each other's expense. Much like a pickup neighborhood soccer game (*pichanga*), the environment of the boardinghouse was informal, competitive, and emphatically male.

While the voices assigned to each character included a variety of regional, cultural, and social types, the boardinghouse milieu mimicked the naturalized heteronormativity of masculine soccer culture.

In *Residencial La Pichanga*, the "gendered soundscape," to use a concept coined by historian Christine Ehrick to depict gender relations in radiophonic settings (2015, 6–7), was a predominantly hetero-masculine phonemic milieu. The few females were submissive and subservient to men and had names that diminished them. The owner of the residencial, Doña Fortunata, was little more than a background figure, always tolerant of her male boarders, the obvious protagonists, and bemused by their antics. Her secondary social status was implied in her name, which linked her to the "feminine" imaginary of canonical literature, namely, the 1887 novel *Fortunata y Jacinta* by Spanish writer Benito Pérez Galdós. In this novel, the female character comes from a poor class but manages to climb the social ladder thanks to her "feminine" beauty (Pérez Galdós 1887). The Fortunata of *Residencial La Pichanga* seems at first to be a "fortunate" woman in a masculinized milieu, but she gets helplessly entangled in soccer culture's web of toxic masculinity. Two other female characters, Doña Fortunata's domestic workers, were equally belittled. Socorro, which means "Help!," was the name given to one boardinghouse maid. The other maid's name, La Popi, was even more derisive and sexist; *popi* in Chilean Spanish slang is a diminutive for a person's ass. In a similar vein, Pirulín, the male cook, suffered the same scornful treatment as the maids, or worse. In the show's heteropatriarchal setting, being a "man" in the supposedly "women's" space of the kitchen was itself stigmatizing; his name and his effeminate voice further situated him as the *marica* (queer) in the show. In the age-old tradition of comedy, characters' names readily indicate their social standing and personality. In this show, character names reflected the gendered imagination of the radio comedy's male authorship. Along with its plotlines of male-dominated gender relations, the show had an undeniably patriarchal imprint, and the vocality of the male characters added a heteronormative cacophony of masculinized voices. Among those voices, Indio Pije's resounded as a particularly abrasive heterosexist male.

Studying the tapes, my auditory memory reconnects me to the broadcasts of *Residencial La Pichanga* that I used to follow at night in the early 1970s. True, I don't hear Indio Pije in the scarce available recordings of the comic radio theater. Indio Pije's absence from the audio tape has a certain situational logic. As the voice of Green Cross, a minor-league team compared to "the big ones" in Chile—like Colo Colo, Universidad de Chile, and Universidad Católica—Indio Pije was not always present in the boardinghouse's gatherings. But I do remember that every time Indio Pije burst onto the airwaves, Ruiz's performance guaranteed that the character entered the comedic competition between "manly" voices by stretching and amplifying the "male" power of his vocal cords. The mnemonic effect of his cartoonishly hyperbolic voice reverberates in my head. My ears transport me back in time: Indio Pije is on the air! He is almost shouting. His voice is never soft, discreet, or overpowered by the others. Stridency is his hook. It is the embodiment of the fictive, assimilated Indian who wants to sound like a Chilean pije.

Through the character's outrageous mannerisms, coarse voice, and volume, Ruiz and César Enrique Rossel, the producer, made Indio Pije into a memorable caricature. All the characters were subject to comic distortion, but the Indian and the homosexual (Pirulín) were easy targets of exaggeration. Like the women on the show, they occupied a secondary position. Indio Pije emphasized the "pije" and the rustic elements of his voice as a tool to define his comic persona and detach himself from his lesser status. Ruiz thus managed to give Indio Pije an outsize presence in the heteropatriarchal setting of macho men who were always competing for attention in the heated, comical soccer rivalry.

INDIO PIJE IN PRINT

The appeal of *revistas de historietas*, or simply *revistas*, grew throughout the twentieth century. These mass-culture publications, averaging between about thirty and forty pages, were inexpensive and accessible. Like their American comic book counterparts, they were priced to be affordable and sold at sidewalk kiosks in cities across the country.[13] In 1965, Rossel, the program's creator, director, and scriptwriter, transported many of the characters from radio to *La Pichanga*, a full-color publication.

Like radio dramas, revistas benefited from the preeminence that *the popular* had acquired in Chilean public discourse after the 1940s. Political movements—right, left, and center—were both cause and effect of the social pressure to incorporate the middle and working classes and low-income sectors into the state's economic, political, educational, cultural, and health programs. It was the era of political populism in

Latin America. This trend gained influence in Argentina under the charismatic leadership of Juan Domingo Perón. In Chile, populism permeated the leadership and political platform of several Chilean governments.[14] Like *Peronismo* in Argentina, Chile's populist politics included periods of right-wing inflected leadership (that of military and political leader Carlos Ibáñez del Campos, 1952–58) but it received the greatest support from the center-left Popular Front from its ascension to power in 1938 to its defeat by Augusto Pinochet in 1973. Contemporaneous with the midcentury's populist politics, Chile's mediascape was undergoing a transformation. Movie theaters became accessible in cities, and radio gained traction throughout the country. Historians Simon Collier and William Sater note that during this period, "horizons expanded" for mass media and communications (1996, 295).[15] The print industry had to adapt rapidly. The revista became a materially and aesthetically fitting genre for this era of political and cultural populism.

By the early 1960s, publishing houses such as Zig Zag and Lord Cochrane had leading roles in the revista industry's growth. Zig Zag expanded its business in 1962 after signing a contract with Walt Disney to carry out the "translation, publication, and distribution of its magazines" in the Southern Cone of Latin America. Lord Cochrane came to a similar arrangement with King Features Syndicate, another US print company (Rojas Flores 2016, 71–78). This was the publishing climate in which Rossel decided to experiment with his characters in the revista format. Publishing the periodic revistas allowed the voice and figure of Indio Pije to take advantage of the new popular culture market for print.

To visualize the radiophonic voices of the *Residencial* cast, Rossel called on reputable illustrators in Chile's print culture. He challenged them to imagine and materialize the radio show's characters in the two-dimensional graphic medium. To make them come alive, the characters had to retain their personal characteristics and idiosyncrasies through a combination of text and graphic techniques that would readily suggest, for example, Indio Pije's pretentious, high-pitched voice. Other notable features and quirks of the boardinghouse radio comedy characters also needed to be translated. What Rossel asked the illustrators to do is best understood as a form of transduction.

As theorized by sound studies scholars such as Jonathan Sterne, much like the use of "transducers" in sound reproduction technologies, transduction involves "turn[ing] sound into something and that something else back into sound" (2003, 22). Sociologist and philosopher of technologies Adrian Mackenzie points out that transduction knots and interconnect "diverse realities," including "corporeal, geographical, economic, conceptual, biopolitical, geopolitical and affective dimensions." For him, "to think transductively is to mediate between different orders, to place heterogeneous realities in contact, and to become something different" (Mackenzie 2002, 18). The transposition of *Residencial La Pichanga*'s characters from the radiophonic to the graphic realm entails the sort of transduction Mackenzie describes; it also invites readers to act as viewers as well as listeners, attentive to the sounds that are encrypted in visual images.

Attention to the intricate crossings between visuality and sonority has become a relevant step for interpreting images in contemporary discussions. In her study on identification photography and similar figurations of Black bodies, Tina Campt proposes "to 'listen to' rather than simply 'look at' images." She engages sound as "an inherently embodied modality constituted by vibration and contact." Through this sensory incitement, sonic dimensions intermingle in "the lower frequencies" of images and perceptually and haptically generate an experience of "felt" sound that, "like hum, resonates in and as vibration" (Campt 2017, 6–7). In a similar vein, art historian Michael Gaudio offers a suggestive intersection of the visual and the sonic. He traces parallels between the experiences of the graphic inscription of sound in literature and music, and the resonances of silent film. For Gaudio, "the way musical notation or the words of a poem" appear on paper "might stimulate an imagined experience of sound." Reflecting on how sound modulates meaning in silent film, he suggests that the process demands that "we put to work 'aural imagination'" (Gaudio 2019, xiii). Based on his study of visual images across time, from "speaking paintings and sculptures" to works of silent film, Gaudio argues, "By putting [images] into conversation with sound, by considering how these pictures put the limits of supposedly silent visual media to the test," it is possible "to bring into focus their engagement with an aural imagination" (Gaudio 2019, xv). Along with the prism of transduction theory, these elaborations on sound and image by Campt and Gaudio helps us interpret the presence or absence of sound markers in the combined graphic and verbal images that Rossel's illustrators used to portray Indio Pije on the page.

Take the full-page panel signed by Tom, the pen name of Chilean comic strip draftsman Víctor Hugo Aguirre (figure 2.2) on the eighteenth page of issue 11 of the *Residencial La Pichanga* revista. The drawing portrays

FIGURE 2.2 · Cartoon comic interchange between Indio Pije and Viejito Magallanes. *Residencial La Pichanga*, no. 11 (unpaginated). Santiago, Chile. 10/20/1965.

Indio Pije sitting on a chair and chatting with the character of La Residencial, who represents the Santiago-based soccer team Magallanes.[16]

The seated Indio Pije wears a cape that covers his soccer attire and most of his body. The cape resembles a Creolized version of a Mapuche poncho, or *makuñ*, even though a large Maltese cross—a Western, Christian symbol—is inscribed on the front of the cape, an important symbol of the Green Cross soccer club. Indio Pije's neatly combed black hair stresses his wingka pije striving, while in contrast his exaggeratedly large head and chubby cheeks makes him look childish. It is the physiognomy of the Indian body as seen through the eyes of a Chilean cartoonist. His Indianness is further marked by a headband that, by its very shape, suggests the traditional trarilonko, even though the design lacks cultural specificity. The wingka bourgeois hairstyle as well as the transmutation of the *makuñ* and trarilonko into a cape and headband of an aesthetically Euro-Chilean design combine to endow Indio Pije with the self-contradictory traits of an assimilated Mapuche. In the center of the panel, we see three small circles and four long feather-like signs, plus a fifth that seems to be a smaller version of the latter, floating around Indio Pije's head. The four larger and longer graphic signs cartoonishly recall the stereotypical image of an "Indian" crowned by a feathered headdress. Furthermore, since the feather-like signs seem to illustrate the loud voice of Indio Pije, they function as signifiers of what Campt calls "vibration" within the process of seeing, feeling, and "listening to images" (2017, 7).

As for the non-Indigenous character in the panel, the Viejito Magallanes (Old Man Magellan), in contrast to the Indian Pije, is a figure of imposing posture and action. The Viejito Magallanes stands. He is a tall, thin, bald man with a slender mustache. His eyeglasses identify him as learned. In contrast to the extravagantly costumed Indio Pije, he wears the basic components of the Magallanes team uniform—jersey, shorts, and cleats. The towering Viejito Magallanes jabs his right index finger at Indio Pije as he looms over the smaller man. The positioning of the two bodies in the panel indicates an asymmetry that is emphasized by the panel's accompanying text:

> VIEJITO MAGALLANES: ¿Qué dirías tú si alguien se sentara en tu sombrero?
>
> INDIO PIJE: ¡Atamalica! ¡Atamalica! Y lo sentaría en la pica.
>
> VIEJITO: Qué bien, porque te estás sentando en mi sombrero.

OLD MAN MAGELLAN: What would you say if someone sat on your hat?

INDIO PIJE: Atamalica! Atamalica! I would seat him on the sharpened stick.

OLD MAN: Excellent, because you are sitting on my hat.

As we read here, the Viejito Magallanes asks, "What would you say if someone sat on your hat?" In his hasty response, Indio Pije refers to the old story of Caupolican's death, the Mapuche warrior who was forced by Spanish colonizers to sit on a sharpened stick. Then, Viejito Magallanes, the Chilean settler character, sarcastically suggests that Indio Pije should suffer the same fate, since he is sitting on the man's hat. The brief dialogue portrays the Indian character as a not very thoughtful person.

The cartoon's graphic and verbal details illustrate the asymmetry of the Indigenous and the Chilean Creole characters. Discussing humor and racial logic in her study of Persian blackface in cartoons, scholar Parisa Vaziri remarks on the "connection between comics and the racial" that "revolves around the reduction of human complexity intrinsic to the process and functioning of both comic form and of racialization." In her view, this reduction entails a "simplification" that, in a sort of "thaumaturgic" manner, "conditions conceptualization, identification and 'closure' in the process of comic meaning making" (Vaziri 2021, n.p.). Such a racialized simplification of the Indian character surfaces in the image and script here. The reductive subalternation of the figure of the Indian unfolds visually and operates verbally. Indeed, the dialogue that accompanies the graphic presents the Chilean figure calmly posing his question and then offering his conclusion in a logical manner. In contrast, Indio Pije is depicted as an immature loudmouth.

The exclamation marks that punctuate his verbal intervention—"¡Atamalica! ¡Atamalica!"—signal the unconstrained energy of his speech. This was characteristic of how Indio Pije entered a scene on the radio show, a verbal tic replicated in the cartoon. A sui generis creation of Indio Pije, "¡Atamalica!" is difficult to translate. Since *Atamalica* ends in *-ica* and rhymes with *pica* (sharpened stick), it might refer to the stick. However, Indio Pije's expression here, as on the radio, registers simply as mere babble. His verbal illegibility marks him with an infantilized, exoticized cultural and "racial Otherness" (Deloria 1998, 132) and confirms the stereotypical Chilean mainstream view that Indios,

FIGURE 2.3 · Cartoon comic drama involving Indio Pije, Pirulín, and a nameless female character. *Residencial La Pichanga*, no. 3 (unpaginated). Santiago, Chile, 1965.

due to their purportedly "uncivilized" and "savage" condition, speak a "rare" but less sophisticated language.

In another issue of the revista, Indio Pije again babbles in indecipherable gibberish, "¡Atamaloto! "¡Atamaloto!" (see figure 2.3). This time Pirulín, the effeminate, queer boardinghouse cook, attempts to explain. When an objectified and sexualized woman character asks him to "translate" what Indio Pije is saying—"¿Qué dice, Don Pirulín?" (What is he saying, Don Pirulín?)—Pirulín responds, also in an exclamatory tone, "¡Dice que lo que más admira en usted es el . . ." (He says that what he admires the most in you is . . .). Certainly, the ellipsis signals that Pirulín is embarrassed and reluctant to complete what Indio Pije just uttered. The woman gets the point but blames the sexist slur not on Indio Pije but on the queer/*marica* character.

The third panel, in the time-honored tradition of a three-panel strip, delivers the punchline. It shows us that Indio Pije remains phallically erect in his hetero-male subject's position, while the queer Pirulín, in contrast, is on the ground, disheveled and holding his head as if he'd been hit. The woman stalks off with her backside still in the frame (and perhaps in the imagination of male viewers). Here again, the only way to understand Indio Pije is to make a critical association with similar ending phonemes in Spanish and graphic details. Indio Pije's illiterate interjection—"¡Atamaloto! "¡Atamaloto!"—may well refer to the woman's *poto* (butt), as, tellingly, that is what is left of the woman's image as she exits.

What we read and see in both the single panel and three-panel comic strip is an unfolding of graphic signs that, to use Gaudio's terms, can "stimulate an imagined experience of sound" and incite "aural imagination" in readers. But, to grasp the cartoon's full graphic and sonic significance, we must pay attention to aural elements (phonemes) and verbal signifiers (punctuation such as ellipsis and exclamation marks), plus the canny graphic details of costuming, the positioning of characters, and the textual suggestions of linguistic savagery. Listening through these visual images, using alkütun, reveals how the radio show's caricature of the shouting Indio is "translated" or transduced into the graphic domain and, ultimately, how acoustic colonialism transits from medium to medium.

INDIO PIJE AS CUMBIA SINGER

In 1975, "El Indio Pije" debuted with an LP titled *Qué pasa en la ruca: Show de cumbias*. Indio Pije thus crossed media once again when comedian Ernesto Ruiz, capitalizing on the character's radio popularity, recorded this album of cumbias in the voice of this well-recognized figure. Cumbia is a genre of danceable music originally from Colombia, with antecedents in Afro and Indigenous musical and dancing traditions from the region (Calle 2021; Ochoa 2016). Ruiz's decision to pair Indio Pije's voice with cumbia was cunning. By the early 1970s, cumbia was by far the most popular dance music in the country. In the early 1960s, cumbia performers and composers had begun taking over the popular music scene in Chile, mainly through radio and urban venues like cabarets and bars (Ardito et al. 2016, 107–8).[17] From the early 1960s to the late 1970s, radio was the most influential and massive medium in Chile; it became the main channel for disseminating cumbias throughout the country.[18] Transporting the voice of Indio Pije to the cumbia musical format amplified its popularity. Indeed, when the six-song LP by Ruiz came out in 1975, as far as I can remember, the title song rose to the top of popular music's radio rankings.

To adapt to cumbia and strengthen the character's performative Indianness, a band called the Sonora Trutruquera accompanied Indio Pije's voice (see figure 2.4). Although the band's name suggested Indigenous origins, with its reference to the *trutruka*, one of the most traditional Mapuche wind instruments, it did not include any Native instruments.

FIGURE 2.4 · Cover of the cumbia songs album *Qué pasa en la ruca: Show de cumbias*, an LP by Ernesto Ruiz as El Indio Pije (1975).

Instead, the band played Western instruments that produced the characteristic sounds of a "tropical orchestra." Heavy on trumpets and electronic keyboards that commonly accompanied *cumbia chilena*, Ruiz's backup band drew on a style popularized by the Chilean group Sonora Palacios, founded in 1962, a few years before the beginning of *Residencial La Pichanga* and the start of Indio Pije's radio performances. Neither the band's past nor its present had Indigenous roots. In its lyrics and in Ruiz's performance style, the album's title song, "Qué pasa en la ruca," projects a White-mediated view of Indigenous life. Here are the beginning and middle sections of the song's lyrics:

> ¡Ayayay! ¡Jaja!
> ¡Alegría, alegría!
> ¡Chita la fiesta!

¿Qué pasa en la ruca?
¿Qué pasa en la ruca?
Indio que entra
me pega en la nuca.
¿Qué pasa en la ruca?
Que Indio que entra
me pega en la nuca.
Ya tengo mi ruca CORVI.
Me la acaban de entregar.
Con un machitun muy grande
yo la quiero inaugurar.
¡Ayayay!
[*Recited*]
Me entregaron mi ruca pu oye.
¡Oi, que alegría!
Ay ay ay! Ha ha!

Joy! Joy!
Good heavens! What a party!
What's happening in the ruca?
What's happening in the ruca?
Indian man who comes in,
he cheats on me.
What's happening in the ruca?
Indian man comes in,
he cheats on me.
I already have my CORVI home.
They just gave it to me.
With a big Mapuche blessing ceremony
I want to open its doors.
Ay ay ay!
[*Recited*]
They gave me my house.
Wow, what a joy!

The song opens with interjections that depict how the settler actor and comedian, through his colonial ear, conceives the voice of an Indigenous person: annoyingly high-pitched and noisily disruptive. The volume is loud, the tone rough. Ruiz replicates the racialized stereotype of the Native's voice as infrahuman, uncontrollable, barbaric, primitive, savage noise. This racial edge of sonic representation is what Stoever, in her analysis of the Black experience in the United States, unveils as the codification of the "sounds of racialized bodies . . . as 'noise,' sound's loud and unruly Other" (2016, 12), which so evidently takes shape in Ruiz's vocal and corporeal performance of Indio Pije.

Through the reiterated uses of interjections and those high-pitched irruptions, Ruiz vocalizes Indio Pije as the character he performed in the radio studio in the 1960s and 1970s. The lyrics and score of "Qué pasa en la ruca" alternate between short phrases, sudden exclamations, and Chilean colloquial expressions, such as *chita* or *pu' oye*. Ruiz aims to depict a stereotyped idea of the Mapuche "Indian," namely, illiterate and with limited command of Spanish. The loudness of his voice conveys the identity and ambitions of a subject who, as a social climber, wants to vocally stand out.

Singing as opposed to acting the role of Indio Pije gives Ruiz access to musical techniques for strengthening the timbre and intensity of his voice to exaggerate the character's manhood (*hombría*). This gendered vocal inflection accompanies the male character's ambition for gaining "superior" class status and patriarchal authority in the Chilean Creole setting. The hypermasculinized vocal quality amplifies the machismo of the lyrics even more. This hypermasculinization is reinforced in the second stanza, the lyrics of which form the chorus of the song. Here Indio Pije refers to his Mapuche wife—his *india*—as an unfaithful woman who cheats on him with every male "Indian" who enters his home: "Indio que entra / me pega en la nuca" (Indian man comes in / he cheats on me). Indio Pije ends by simply echoing the White male settler cliché that Indigenous women are "easy women," with its latent implication that the Indigenous woman is a mere object of male manipulation, passive and voiceless before an "Indian" social scene of rivalry, distrust, and betrayal among men. Added to this, the fictive Indian "fiesta" imagined by Ruiz also exposes Mapuche men as equally untrustworthy and lacking the moral values that Western settler societies supposedly cultivate, such as (male) honor. What Spanish feminist philosopher Celia Amorós calls "patriarchal pacts between males" (2007, 119) takes a degraded shape in the lyrics of "Qué pasa en la ruca." In Ruiz's cumbia song, machismo and racism cross-fertilize as part of a twisted colonial narrative. Through Indio Pije, the Chilean Creole comedian manages to vocalize and project a cartoonish and degrading portrayal of Mapuche masculinity. His song constitutes a sort of twentieth-century echo of the final scenes of Alberto Blest Gana's novel *Mariluán* (1862), already discussed in chapter 1, that is, the settler disfigurement of Mapuche life as a social setting untrustworthy, prone to disloyalty, betrayal, and lacking in morals.

If the character's most problematic features speak to his misleading representations of the Mapuche, his "pije" side is depicted somewhat more positively, especially his accomplishments as a Mapuche migrant in the city. Right after the song's chorus, Indio Pije tells us how happy he is to have become a first-time homeowner, thanks to the Chilean state's government-housing program, Corporación de la Vivienda, or CORVI: "Ya tengo mi ruca corvi / me la acaban de entregar" (I have my CORVI home now / they just gave it to me). Driven by a politics of "social housing" that aimed "to solve the needs of lower-income social groups" (Aguirre and Rabí 2009), CORVI was instrumental in providing affordable housing not only to the low-income working class but also to middle-class residents of the greater Santiago area and the main urban centers in the provinces from the 1950s to the early 1970s.[19]

Soon after the coup, Pinochet abandoned the welfare-state model. By 1975, he had embraced the free-market ideology of Milton Friedman, the Chicago school of economics, and their local disciples from Catholic University of Chile. This monetarist approach paved the way for neoliberal policy on housing and urban development, upending or simply dismantling state-sponsored social housing policies and related government initiatives, such as CORVI, inherited from the previous era of populism and socialist reformism in Chile (Vergara-Perucich and Boano 2021, 198–99). It seems paradoxical for Ruiz's Indian persona to celebrate CORVI amid its progressive weakening under Pinochet's regime; in any case, what is clear, in post-1973 Chile, is that the assimilated figure of Indio Pije is making explicit his attachment to the authoritarian state's benevolence.

Yet the idea of an assimilated Mapuche also coexisted with the era of the Chilean welfare state. Like CORVI, founded in 1953, Indio Pije came to life in the mid-1950s. Furthermore, *Residencial La Pichanga* was launched in 1965, when the government of Chile instituted the Ministry of Housing and Urbanism to support the role of CORVI in "materializing housing policy" and respond more effectively to "a moment of strong social participation with the massive incorporation of diverse social groups into the housing policies" of the state (Aguirre and Rabí 2009). This parallel between the state's policies of social integration and the radioteatro's assimilationism of the Indigenous subject reflects a logic that was characteristics of populist politics in Chile and throughout Latin America in the 1950s, 1960s, and 1970s. In fact, in the mindset of those who adhered to Chilean populism, Indio Pije embodies the conflation of social and symbolic qualities that an assimilated Native would possess.

The next verses of "Qué pasa en la ruca" move to the problematic verbalization of Indio Pije's gender views. In his musical caricature, Ruiz employs sexist expressions such as "Mi india tiene buen cuero" (My india is hot meat) or "Chita mi india es güena, oye" (Shoot, my india is hot, man) to make it clear the Indigenous male treats women like sexual objects, while at the same time he views Mapuche men as drunk and dirty. More racializing stigmas end the song:

¡Chita la fiesta pa' güena!
¡Quién se lo iba a imaginar!
Me levantaron la india.
¡Ahora me largo a chupar!
[*Recited*]
¡Ayayay! ¡Jajay!
Me levantaron la india estos mugrientos, oye.

Wow, what a good party!
Who would have imagined!
They stole my Indian woman!
Now I am gonna go get drunk!
[*Recited*]
Ay ay ay! Ha ha!
Hey, these dirty injuns stole my Indian woman.

Rather than sung, this last stanza of the cumbia, like the initial segment, is spoken in the peculiar radiophonic tones of Indio Pije, with abundance of interjections and an annoying, abrasive, and loud voice frequency. A series of interjections in the final verses reinforce all these features. The tone of the Chilean "pije"—that is, the typical voice of a macho—ends the song. Referring to the still popular *Residencial La Pichanga* radio show, with its heteropatriarchal soccer culture and its scenario of magnified "masculine domination" (Bourdieu [1998] 2001), the last part of the song depicts the assimilation of the Indian character into the vocal inflections of Chilean "pije" masculinity through linguistic turns of speech. Ruiz / Indio Pije uses Chilean slang expressions that were common among men: "Chita la fiesta pa' güena" (What a good party!) or "chupar" (literarily "to suck," but here to drink alcohol). The song ends by invoking tropes of drunkenness and dirtiness that the White settler imagination associates with the Mapuche lifestyle. The last stanza of "Qué pasa en la ruca" insists on a racialized stereotype that Mapuche men can't hold their liquor. Even as Indio Pije decides to get drunk himself (*me largo a chupar*), he deems his Indigenous peers as "dirty" (*estos mugrientos*). Ruiz draws on the combined effects of the lyrics and cumbia music to mock his Indigenous subject and calls on the stereotypes of the drunk and dirty Indian that would resonate with his audience. False Indian though he may be, here Indio Pije is festive, dancing, and once again the butt of the joke.

MOVING ON

The musical and radiophonic impersonation of the Mapuche, involving Ruiz's throat, voice, and body, entailed a racialized staging of indigeneity in Chile. The voice that Ruiz adopted sonically supplanted and disfigured the Indigenous subject, to produce an "Indian" who came to occupy a niche in Chilean comic radio theater and popular music. Ruiz's impersonation of his oxymoronic character—Indio Pije—is the radiophonic parallel to Blest Gana's literary mode of colonial mediation and disfigurement of the Mapuche. I situate both Blest Gana's textual narrative and Ernesto Ruiz's radio-theatrical and musical performance within the broad realm of acoustic colonialism—the sonic and auditory practices through which the voices of the Mapuche people have become supplanted, mediated, disfigured, and dispossessed of public audibility.

Historically, this process was spawned by the assimilationist policies of Chilean state agents and capitalist forces toward Indigenous Peoples. The same Chilean Creole framing of Mapuche representations that epitomizes the semantic dimensions of Blest Gana's text and Ruiz's multigenre performance is common in the broader sonic domain. These representations generate and reinforce an acoustic, cultural, and historical distortion of the Mapuche, underscored by the same structural power relations that foster racism, male heteronormative sexism, and colonialism in Chilean society. Misrepresentations of the voices and even of the silences of the Mapuche form part of the machinery put in motion to perpetuate the hegemony of the Chilean nation-state. Since the nineteenth century, Creole agency over the voice and representation of the Mapuche has been a generalized norm in the Chilean mediascape. The colonial ear has enjoyed a privileged position that filters and then represents (more accurately, misrepresents) "Indians" or "Araucanians," paving the way for the replacement and sonic dispossession of Indigenous voices dismissing and supplanting the actual Mapuche.

Ruiz's simulation of an Indigenous character is a corporeal and racial performance of colonial settlement in the "acoustic territory" (LaBelle 2010). The way Ruiz directly and unashamedly jokes with the figuration, or disfigurement, of a supposedly Indigenous person is a Chilean version of the North American settler tradition of "playing Indian" (Deloria 1998), or a parallel to the theatrical and cinematic tradition of "redface" (Raheja 2010). The staging of Indio Pije in radio, music, and revista was a form of occupation and colonization. The comic possession and performance of this fictive yet corporealized and figurative indigeneity—first by the Green Cross soccer team, then by Rossel, the scriptwriter, and then Ruiz—echoes what scholar Lorenzo Veracini defines as "transfer by performance." In this tactic of identity invasion and dispossession, "As settlers *occupy* native identities, indigenous people are transferred away" (2010, 47). This takes place, for example, when settlers "dress up as natives," as Deloria conceptualizes; the practice displaces the actual Indigenous bodies and embodies an "elimination of the native" (Wolfe 2006). This bodily occupation and (dis)possession took place through the caricatured staging of a made-up Mapuche.

....

With Indio Pije's voice still resonating in my auditory memory, I return to the scene of my childhood. It is the early 1970s and I am back in my grandmother María Clara Lefno Huechante's house. My memory takes me back to one of those nights at home, listening to the radio. When the radio drama broadcast ends, I get to take control of the dial. As I search for nighttime shows, I shift the frequency from AM *to shortwave. I start hearing voices from other countries of South America. Voices in Spanish. Voices in Portuguese. Voices in English. It is close to midnight. Here in the rural areas of Füta Willi Mapu (Great Lands of the South), it is also possible to capture radio transmissions from the other side, the east side, of the Andean Cordillera, that is, Puel Mapu (Lands of the East) or what is currently called Argentina. While exploring the radio dial, I accidentally tune in to a radio show in Mapudungun, the Mapuche language. It is unclear to me whether the broadcaster is an Argentine Creole or a Mapuche, but I listen attentively. The broadcaster, giving a lesson on counting, repeats the numbers in Mapudungun again and again. It awakens something in me. As a part of a Mapuche-Williche community that was prohibited from speaking our native language by the still-influential Catholic missionization, I rarely heard Mapudungun. Occasionally I passed by the home of elder Reynaldo Martin, a neighbor who used to chant in Mapudungun at nightfall around his kitchen hearth. Listening now to some words in Mapudungun from a radio station is an acoustic awakening for me. I look for a notebook and a pen. I start writing down what I can catch while the broadcaster says, "Kiñe, epu, küla, meli, kechu, kayu, regle, pura, aylla, mari," repeating the sequence of numbers from one to ten in Mapudungun. The aurality of Indio Pije now resounds from afar. The words in Mapudungun seem to interfere or conflict with the radiophonic and phonetic script of my linguistically and acoustically colonized Williche ears, but those native words intrigue me and seem full of promise.*

....

My childhood experience of counter-listening within a colonized sound environment was common for the many Mapuche who turned on their transistor radios in search of an alternative voicing of Ngulu Mapu. The radiophonic airwaves and music industry were entangled in the existing power relations in society. During most of the twentieth century, radio staged a politics of voice that served the racial, cultural, and social imaginary of the Chilean nation-state as a colonizing agent. By the early 1970s, an open struggle by the Mapuche for vocal autonomy and agency in the hegemonic mediascape had not yet begun. However, in the 1980s and early 1990s, the emergence first of Mapuche literary voices and then a similar current in the acoustic territory of radio, would interject a challenge in the script of acoustic colonialism. The following chapters focus on the response to the Chilean Creole politics of historical misrepresentation and disfigurement of the Mapuche. A countercurrent to acoustic colonialism, these creative modes of asserting our own forms of audibility and agency in the domain of sound became critical forms of Indigenous interference.

THREE

....

Listening Poetically: A Land That Resounds and Sings

Something unusual was afoot in Santiago in 1989, aside from, though possibly related to, the prodemocracy grassroots mobilization against the Pinochet regime in which, as a recent college graduate, I was engaged. After hearing an announcement in a newspaper, I went to the headquarters of Editorial Universitaria, in the Commune of Providencia, to attend the launch of Leonel Lienlaf's first published book of poetry, Se ha despertado el ave de mi corazón *(The Bird of My Heart Has Awakened). The event's room was so packed I could not get in. A few days later, however, when I was able to buy and read the book, it awakened something in my own heart as a young Mapuche-Williche living in Santiago, away from my native land. In fact, the publication and launch of this book of poems marked a significant critical disruption in the catalog of the "colonial library" (Mudimbe 1994) in Chile. The bird that awakens in Lienlaf's heart was perhaps a synecdoche of a will to liberate the colonized voices, bodies, and territories of Ngulu Mapu. Between the 1980s and 1990s, a series of literary works by Mapuche authors emerged as Indigenous interferences in literature. One exceptional publication spawned others, creating a Mapuche catalog. Lienlaf's 1989 book was the first to incorporate*

Mapudungun as the primary language in a book of Mapuche authorship. Like other poetry readers, since then I could not only "see" the Mapuche language printed on the surface of paper but also had the opportunity of "listening to" Mapudungun in literary events. Indeed, in the late 1990s, in one of the reading rooms of the National Library of Chile in Santiago, Leonel Lienlaf started his recital singing, in Mapudungun, the third part of his poem titled "Zokiñmangey ñi furi" from the book Se ha despertado el ave de mi corazón. *His voice transported the text from the printed page to the acoustic realm of the chant. This vocal act gave new life to his Mapudungun-written poem through the timbre and intonations of his throat and voicing, unexpectedly transmuting his identity as "poet" to ülkantuchefe—a chanter, the one who sings in Mapuche community life. Then, following this performance, Lienlaf "read" the version of the same poem in Spanish. Through these poetic and sonic gestures, Lienlaf vocally interfered with the customary script of poetry readings and disturbed the library's aural setting by positioning Mapudungun as a literary language and challenging the colonial supremacy of Spanish in Chile's print culture and mediascape.*

....

The National Library (Biblioteca Nacional) was founded on August 19, 1813, in Santiago, the capital city of the newly formed "Republic of Chile," three years after the Chilean nation-state itself had been founded on the scaffolding of the previous Spanish colonial holding (Collier and Sater 1996, 32–50). To Mapuche thinking, the wingka people—first the Spaniards and then the Chileans—occupied their territory not only in economic, social, and political terms but also linguistically and culturally. Under wingka leadership, the government of the Republic of Chile was set up in Santiago, a colonial urban settlement established in the territory of the already removed Indigenous Pikunche (People of the North).

A century later, the National Library settled where it remains today, in a neoclassical concrete building at 651 Alameda Avenue in the heart of the Chilean capital, an iconic space for Creole print culture, in which Spanish functioned as a key component of the colonizing acoustics of the emerging nation-state. Erected as a symbolic and material patrimony of Creole settler culture, the National Library served as the repository of important private collections of the lettered elites as well as the myriads of books published from the early nineteenth century onward. By the second half of the 1800s, Chilean written literature had become a mechanism of colonial mediation of the voices of Ngulu Mapu. Then, over the twentieth century, the radiophonic medium emerged as another agent of the same settler power's articulation. The Chilean hegemonic mediascape thus helped strengthen the deployment of acoustic colonialism on the territories and ears of the Mapuche.

Though notable, Lienlaf's public reading in Mapudungun/Spanish at the National Library of Chile was far from an isolated phenomenon. The event was part of the Mapuche literary movement that was beginning to gain a public presence in Chile in the late 1980s.[1] Elicura Chihuailaf and Lorenzo Aillapan Cayuleo also emerged as Mapuche poets whose writing and public presentations first positioned Mapudungun as a literary language. These poets put the language-territory nexus in writing and vocalized it through the expressive and performative force of Mapudungun, while providing accompanying Spanish versions of their texts. In their works, the acoustic ecology of the Mapuche linguistic, cultural, and territorial imaginary as well as the dissonant omnipresence of colonial history sound in both figurative and literal ways. The appearance of Mapuche poets beginning in the 1980s ran against the grain of a colonizing society and the ideological institution of poetry—namely, "Chilean poetry"—challenging its predominantly monolingual canon.

Spanish prevails in public and private spaces in Chile. Most of the mainstream population has no basic knowledge of the Mapuche language, while English is increasingly influencing public culture "even though few people understand it, and very few speak it" (Zúñiga 2010, 41). In *La lengua Mapuche en el siglo XXI* (2010), Mapuche scholar María Catrileo provides data about the status of Mapudungun as a native language that has been relegated to the margins by the settler society. Referencing the 2006 census in Chile, which identified 604,349 people as Mapuche, she suggests that "between 300,000 and 350,000 Mapuche people still speak Mapudungun with varying degrees of fluency in rural and urban areas between the Metropolitan Region and the Tenth Region" (Catrileo 2010, 38).[2] Catrileo adds that, in their uneven fluency, the population of Mapudungun speakers fluctuated from those who can deliver "a few expressions, such as greetings" to those who practice "a well-crafted speech" (2010, 41). These studies describe the precarious situation of Mapudungun at the beginning of the twenty-first century, as a language displaced to the margins by Chilean settler society. They also offer a telling depiction of the hegemonic linguistic, graphic, and acoustic regime of language within the Chilean nation-state, and help to grasp the dominance of monolingualism in the realms of print culture and mass media in Chile, a sociolinguistic climate emblematic of the deployment of acoustic colonialism over Ngulu Mapu.

For Mapuche poets to make assertive use of Mapudungun in this monolingual context is an assertion of resistance and agency. Their linguistic registers carry out a creative graphic, sonic, literary, aesthetic, and political reterritorialization of the language of the Mapuche. By employing Mapudungun as the primary language, and demoting Spanish versions to supplements, authors such as Lienlaf, Aillapan Cayuleo, and Chihuailaf interfere in the waves of acoustic colonialism. Moreover, by writing in Mapudungun, their works make a significant contribution to the broader Mapuche movement, which aims to counteract centuries of misrepresentation and to reposition the autonomous sounds and voices of Ngulu Mapu and *Wallmapu*, all the lands of the Mapuche Country, more broadly.

This chapter focuses on how Leonel Lienlaf and Lorenzo Aillapan Cayuleo interweave language, sound, and land in their poetry, and how they stage the voices, sounds, and images of their Mapuche-Lafkenche territories through poetic and performative registers. All these dimensions take on multifaceted prominence. The very name Mapudungun juxtaposes and interweaves two realms: *Mapu* (land/earth, territory, environment, universe) and *dungun* (tongue, language, voice, sound, sense). In tying language to earth, Mapudungun diverges from an anthropocentric logic and expresses a linguistic territory of multiple resonances: the phonetics of a world populated by beings that whisper, mumble, talk, shout, sing, weep, scream, whistle, trill, bleat, and moo. These sounds of the nonhuman and human environment are no longer distorted by a colonial ear that hears them as expressions of savagery or targets of mockery, as exemplified in the portrayal of Mapuche cheers as *chivateo* (goat bleating) in Blest Gana's novel *Mariluán* (1862) or the caricatured Mapuche voice in Indio Pije. Instead, the poetry of Lienlaf and Aillapan Cayuleo invite us to engage the acoustic ecology of Ngulu Mapu through Mapuche poetics and linguistics. To do this we must put allkütun into practice, sharpening our ears to engage an Indigenous acoustic ecology imbued with language, voice, body, and territory. We must take up their invitation to exercise literal and figurative modes of attentive listening, that is, to listen poetically to the human and nonhuman, the tangible and nontangible lives of the Mapu.

Figurative modes of "hearing" and "listening" are critical to the process of poetic reading and interpreting. Lienlaf's and Aillapan Cayuleo's poems incite an imaginary listening. Their sound resonates in the ears (physically and mentally) of those who read them. Scholar Marjorie Perloff observes that poetry "inherently involves the structuring of sound," reminding us of Roman Jakobson's view that poetry is "a province where the internal link between sound and meaning changes from latent to patent and manifests itself most palpably and intensely" (2009, 1). Reflecting on Western lyrics and sound in her book *Poetry and the Fate of the Senses* (2002), poet and scholar Susan Stewart reminds us that "poems compel attention to aspects of rhythm, rhyme, consonance, assonance, onomatopoeia, and other forms and patterns of sound" (68). She acknowledges that the "lyric is not music," even though "it bears a history of relation to music" and that writing "has no sound" (68). However, Stewart also offers the idea that poems can textually generate "aural connections" (73). Moving from text to reader, she further theorizes that "the sounds of a poem . . . are heard within a memory of hearing that is the total auditory experience of the listener" (75)—or, if may I add, of the reader as listener.

Stewart's considerations, though insightful, are limited to the "human" domain. Literary scholar Francine Masiello amplifies this perspective, suggesting that in reading a poem, "the spectrum of dimmed voices becomes not only audible but palpable," activating a sensory experience that invites us "to meet and connect with the environment." The Mapuche concept of allkütun captures well the "attention" that Stewart associates with what poems incite, and it adds to Masiello's broader openness to "the environment." This approach entails transforming the practice of close reading into "close listening," but not only in a literal dimension, as critic Charles Bernstein formulates. Allkütun is also an invitation to exercise close and attentive listening to poetic writing and its vocal and sonic reverberations, creatively interweaving both the aural and the textual. Here I borrow Anthony Reed's approach to the intersection of music and poetry in Black arts. For him, "listening must contend with sound neither as pure presence nor as pure symbolic transcendence but as *text*, a structure weave of the phono- and typographic, the grammatical, and the semantic" (A. Reed 2021, 4).

Based on these sonic and literary premises, in this chapter I contend that language, the land, and the universe become resounding territories in the poetry of Lienlaf and Aillapan Cayuleo. This enables them to poetically articulate a semantic, sonorous, and sensorial relationship to Ngulu Mapu. From this listening angle, I discuss the first two collections of poems published by Leonel Lienlaf, *Se ha despertado el ave de mi corazón* (1989) and *Pewma dungun / Palabras soñadas* (Dreamed Words) (2005), and the book of poems titled *Üñümche: Hombre pájaro* (Bird Man) by Lorenzo Aillapan Cayuleo, a corpus that he partially published in Cuba in 1992 and then published in its completed version in Chile in 2003.

Both authors' poems acquire singular resonance in Mapudungun. They are meant to be performed as chants (Lienlaf) or actively vocalized as onomatopoeias (Aillapan Cayuleo), thus transforming the poets' literary practice into verbal and sonic acts. Without losing sight of the relevant graphic and visual dimensions of Lienlaf's and Aillapan Cayuleo's writings, attention to their literal and figurative ways of positioning sound in language is central to my analysis. The following pages offer a way of reading-as-listening to the voices of the *Mapu* and the *Che* in the poetic creations of Lienlaf and Aillapan Cayuleo.

THE DREAM OF THE LAND THAT SHOUTS

Even before Leonel Lienlaf altered the poetry reading habits at the National Library, he broke from wingka language dominance by publishing his first book of poems primarily in Mapudungun, with accompanying "versions" in Spanish. The publication of *Se ha despertado el ave de mi corazón* in 1989 by Editorial Universitaria, the publishing house of the University of Chile in Santiago, represented a pivotal moment in the history of literature in Chile. Prior to 1989, the Mapuche language had never been at the center of a literary work published by a nationally prominent press. The book became a literary tour de force, winning the City of Santiago's Municipal Literary Prize in 1990. The timing of the book's publication, coinciding with a resurgence of Mapuche social and political mobilization and Chile's transition to democracy, symbolically magnified its impact as a linguistic and literary event. Throughout the 1990s and 2000s, Lienlaf expanded his creative work. In 2003, he published his second book of poetry, *Pewma dungu / Palabras soñadas*, also in Mapudungun and Spanish.

Lienlaf was born in 1969 in Alepue, a Lof (community) that is part of the region the Mapuche identify as Lafkenmapu (Lands of the Coast). Lafkenmapu is the region to which the Lafkenche (People of the Coast) belong. It spans the coastal side of the whole Araucanía region down to the seaside of the provinces of Valdivia and Osorno, including the Chiloé Island, in southern Chile. Alepue is on mountainous lands facing the Pacific Ocean, west of the town of San José de la Mariquina, on the northern coast of Valdivia Province. There Lienlaf attended primary school to advance his formal education in Spanish. After completing his sixth year of basic education in Alepue in 1980, Lienlaf moved to a Catholic school in Padre de las Casas, a town south of Temuco, where he continued his secondary education and began to write poetry.

His first two books feature the Mapuche trope of *pewma* (dream) at the "heart" of his poetic language and vision.[3] This trope becomes ubiquitous in his representation of the multiple bodies and spirits of the land, the waters, and the territory; that is, the tangible and intangible forms of life that inhabit the Mapu. Existing studies of these books have tended to interpret the omnipresence of the pewma as part of a dematerialized worldview, exoticized as a mere cosmology, referred to in Spanish as *cosmovisionismo*. In their ahistorical, culturalist, and metaphysical approach, these readings delink the symbolic layers of Lienlaf's poetry from its historical entanglements.[4] In contrast, I approach *Se ha despertado el ave de mi corazón* as an exploration of the poetic and cognitive powers of dreams not only in relation to the intangible forces that underlie Mapuche modes of being and knowing but also as voyages into the materiality of history.

Several poems in Lienlaf's first book link the revelatory powers of dreams to historically loaded images and voices that, in allegorical and referential ways, document a colonial history from the violence of the pre-1800s Spanish rule to that of the later Chilean Creole settlement in Mapuche territories. In *Pewma dungu / Palabras soñadas*, Lienlaf again delves into the juxtapositions of dream, history, and representation. Moreover, this corpus concludes with an allegory of the late twentieth- and early twenty-first-century invasion of the lives of Ngulu Mapu by the depredatory and extractivist forestry industry. Lienlaf's first two books thus reveal the clashes of successive wingka colonialisms with a Mapuche spirit attached to the land as a locus of the dream and the chant.

The land dreams and sings. And it also trembles and sobs. Lienlaf's poetry juxtaposes and intertwines land, language, sound, and history. This is articulated through his poetic and sonic use of Mapudungun as "language" and "voice" of the earth; many of his poems are conceived not only as written artifacts but also as chants to be performed in the native language. In the cultural climate of the 1980s and 1990s, Mapudungun was scorned or simply ignored by mainstream Chileans, while among Mapuche who knew the language, many were embarrassed or afraid to speak it in public or even at home. For native speakers of Mapudungun for whom Spanish was merely a school language, their "Indian" accents and vocal inflections in Spanish were often objects of mockery or caricature in mainstream media, as we saw with Indio Pije in the 1960s, 1970s, and 1980s. By publishing and delivering his work in bilingual editions, Lienlaf legitimized the Mapudungun in the realm of written poetry, giving it prestige in the official literary milieu and public culture of Chile.

Lienlaf grew up in close contact with Mapudungun, spending time at the home of his maternal grandmother, Marcelina Pichun, who told and sang stories around the *kütralwe* (hearth). Relatives of the Lienlaf kinship and members of other Mapuche-Lafkenche communities visited her frequently. The elders had long discussions in Mapudungun. Describing his own experience in this setting, he says, "More than conversation, it was listening; more learning. . . . This was my best school, having to listen. . . . From when I was two until I was ten, I learned how to listen, because that is a fundamental matter: learning to listen and learning how to observe" (Lienlaf, interview by author, 2015).

One of the opening poems of *Se ha despertado el ave de mi corazón* foregrounds the notion of allkütun. This poem is the last of a series titled "Cholkiñmangey" / "Le sacaron la piel" (They Took Off His Skin), which evokes the killing of Lautaro, one of the main Mapuche warrior leaders during the era of Spanish colonialism. The poem's opening scene makes us hear the "screams" and "whistles" of Mapuche people fighting against wingka forces of "death" and destruction. Amid these historical yet vividly remembered scenes of genocide and death, the speaker conveys his sorrow and invites readers to listen:

Rupa-rupangey tripantü,
rupa-rupangey Mapu,
kanchalen ka dewma

pepi dunguwelan.
Allkütumuchi ka puen pipingen. (Lienlaf 1989, 36)

Están pasando los años,
están pasando los nidos sobre el fuego,
está pasando la tierra
y ya me estoy perdiendo
entre las palabras.
Escuchen hablar a mis lágrimas. (Lienlaf 1989, 37)

The years are passing,
the nests over the fire are passing,
the earth is passing
and I am already getting lost
among the words.
Listen to my tears speak.

Lienlaf's poetry challenges us to open our ears and pay attention to what his "tears" say in Mapudungun and then in Spanish: "Allkütumuchi ka puen pipingen" / "Escuchen hablar a mis lágrimas." This poem conveys the angst of a speaker who struggles "between the words," situated between Mapuche voices and experiences of colonial violence. An inexorable passing of time threatens to erase the sense of memory and history. As readers we are invited to empathize through an act of listening that connects us to this history of "tears," to the beleaguered history of the Mapuche, and to the expressive struggle of a self that becomes entangled in language.

In a Chile ruled by Pinochet's dictatorship, Lienlaf imagined a disruptive literary act: publishing a bilingual Mapudungun-Spanish book of poetry. Thus, by the early 1980s, he started drafting poems and drawings that, by the end of the decade, became the basis for *Se ha despertado el ave de mi corazón*. The book invites readers to listen attentively, to navigate the texts in Mapudungun that function as parallels to their Spanish versions. The latter do not work exactly as translations but as transmutations in both content and form. The poet's invitation to exercise attentive listening requires audiences to read and listen across languages, sounds, and histories.

"The bird of my heart" invoked in the title of this seminal book metaphorically embodies a resurgent Mapuche linguistic and cultural identity. The poet from the Alepue community performs the awakening of a non-Chilean notion of poetry: the tradition of ül, or chant, the Mapuche art of linguistic performance and singing that involves a strong linkage between the self and the spirits of earth. This imaginary takes shape in Lienlaf's writing through a constant engagement with the dreams and the voices of the land, as the opening lines of his first book suggest.

> MAPU ÑI PEUMA
> WIRAÜMEKEY
> ÑI PIUKEMEU
>
> EL SUEÑO DE LA TIERRA
> GRITA
> EN MI CORAZÓN . . . (Lienlaf 1989, 29)
>
> THE DREAM OF THE LAND
> SHOUTS
> IN MY HEART . . .

The land here dreams. It is a dream that comes from the very heart of the poem's speaker. Lienlaf uses the typography on the page as a territory, graphically reinforcing the commotion of land and self that resonates in Mapudungun. Placing the Mapudungun text in capital letters above the Spanish version, Lienlaf positions the fundamental place of the land as a living being at the very "heart" of the poetic persona. The land dreams and sounds as an active agent. The shout of the land (represented in the uppercase type) brings both strength and tension to this bilingual compressed text that remains unfinished in the Spanish version. The ellipses perhaps mark the incompleteness of the Mapuche expression in the colonial language.

RESONANCES OF TERRITORIAL STRUGGLE

Lienlaf's poems resonate as the "shout" of Mapuche political struggles for the land. The publication of his first book coincided with a movement for democracy and the terminal phase of the dictatorship. The democratic transition, carried out within the framework of a neoliberal capitalist model, resulted from negotiations between military commanders, the entrepreneurial class, and the political elite. Many Mapuche communities began to rebuild their political and social organizations while denouncing the alliance of the Chilean state with private corporations that promoted and financed the expansion of the forestry industry in Mapuche lands. Against these new invasive forces, land politics emerged as a critical component of the Mapuche movement in the late 1980s and early 1990s. This sense of territorial articulation is confirmed in the very name of the influential Mapuche political organizations founded in Temuco: Aukiñ Wallmapu Ngülam / Consejo de Todas las Tierras (Council of All Lands). Established in 1989 with a focus on politics of self-determination and autonomy, the organization challenged the colonial framework of the Chilean nation by waving the flag of Wallmapu, the Mapuche Nation, a geopolitical space that traverses the Argentina-Chile border. Published in the same year, Lienlaf's *Se ha despertado el ave de mi corazón* was recognized as an intrinsic part of this historical shift, as a "shout" or validation of the Mapuche resurgence.

Lienlaf's poetic shout of "the dream of the land" converges with this momentum by linking politics to a poetics of the Mapu—the territory, the earth, the universe. In his poems, the land is more than an externality. Symbolically and sonorously, Lienlaf expresses how land is internalized in language and subjectivity. "The dream of the land / shouts / in my heart." The poetic persona moves through historical traumas of colonial border violence, through the Mapuche encounter with invasive forces in a geopolitical space that the wingka named La Frontera (The Frontier). In "Rupamum" / "Pasos sobre tu rostro" (Steps on Your Face), Lienlaf underscores this colonial borderland history:

> ina pen kiñe cruz katrünmaetew ñi lonko
> ka kiñe espada bendecipeetew petu ñi lanon.
> Güypechymutrungen
> mi rukamew, ñuke. (Lienlaf 1989, 54)

> y vi una cruz que me cortaba la
> cabeza
> y vi una espada que me bendecía
> antes de mi muerte.
> Soy el tronco, madre
> el que arde
> en el fuego de nuestra ruka. (Lienlaf 1989, 55)

and I saw a cross that cut my
head
and I saw a sword that blessed me
before my death.
I am the trunk, mother
the one which burns
in the hearth of our ruka.

These verses make a reference to the symbolic and historical violence of Christian-Spanish colonization that is more than a mere reminder of the past. It cracks the body of the text in Mapudungun. The words *cruz* (cross), *espada* (sword), and *bendecía* (blessed) are a gash in the phonetic and semantic horizon of the Native language. Linguistic ecology mutates; the language of the land is altered; its singing is fractured. Lienlaf touches on this tear again in "Trig metawe" / "Cántaro trizado" (Cracked Pitcher), one of the final poems of *Se ha despertado*.

Dewma afmekey
tüfachi kelü metawe
trigy deuma.
Mapunmew umgtuay,
kiñe antü
ka wüdüfe
deumatuaeyew. (Lienlaf 1989, 110)

Ya se está acabando
este cántaro
rojo.
Ya se ha trizado
y dormirá
entre las cosas de
la tierra
hasta que un día
otro alfarero
lo reconstruya. (Lienlaf 1989, 111)

It is already ending
this red

pitcher.
It has already been cracked
and it will sleep
among the things of
the earth
until one day
another potter
restores it.

The *kelü metawe* (red pitcher) has become the Mapuche ül. This mutation echoes more strongly in the accompanying Spanish text since *cántaro* (pitcher) is phonetically close to *canto* (song). Thus, the image of the "red pitcher" represents the Mapuche chant that, in its trajectory from page to page, ends up cracked.

THE LANGUAGE OF GALLOPING THE LAND

Lienlaf confers on print culture a notion of poetry that exceeds the linguistic and cultural parameters of the institutional space of "Chilean literature." His verses flow and fly beyond the boundaries of writing. Indeed, many of his poems are meant to be recited and sung. Through public performance, they become chants, reviving the Mapuche poetic and musical genre ül. A poem dedicated to one of Lienlaf's ancestors is read and performed as such, a chant:

Kiñe wentru lelfünmew
wiraf-wirafngey,
ñichiripa ka wirafküley
wente ñi kawellu
mütrümpelu ñi trewa.
Kom Mapu
allkuy ñi dungu.
Kedintu-kedintu
Lautraro-Lautraro
pi ñi wirafün,
ñi pu trewa inanieyew
kürüfreke. (Lienlaf 1989, 34)

Un hombre va galopando
en la pampa
y su chiripa galopa
sobre su caballo
llamando a sus perros.
Toda la pampa escucha
sus gritos
Kedintu-Kedintu
Lautraro-Lautraro
dice su galopar.
y sus perros lo siguen
como el viento. (Lienlaf 1989, 35)[5]

A man goes galloping
across the prairie
and his chiripa gallops
on his horse
calling up his dogs.
The whole prairie listens
to his shouts
Kedintu-Kedintu
Lautraro-Lautraro
his gallop says
and his dogs follow him
like the wind.

What "gallops" here is the long history of Mapuche struggle. The human subject, the Mapuche, or "person from the earth," screams, and the pampa listens. The poem-chant evokes a Mapuche man ("Kiñe wentru . . .") on his horse, an image of enhancement and empowerment. According to the poet, this poem primarily evokes an ancestor of his, "whose name was Lautraro, the grandfather of my maternal grandmother" (Lienlaf, interview by author, 2015). Lienlaf recalls that Lautraro participated in the Mapuche uprisings against the invasion by the Chilean military and colonizers in the second half of the nineteenth century. In this context, Lienlaf also evokes the historic presence of Chilean military in Mapuche territories as the Creole nation-state planned to take control of Native lands and consolidate its settler colonial geopolitics by that time. This was the so-called Pacification Campaign led by Colonel Cornelio Saavedra, the Chilean military enterprise that was launched with the occupation of the town of Angol (Province of Malleco) in 1862 and ended with the arrival of Chilean troops in the old colonial settlement of Villarrica (province of Cautín) in 1883.[6] In Lienlaf's family memory, Lautraro was among the many Mapuche-Lafkenche from Willi Mapu (Lands of the South) who responded to the calls of resisting communities from the Araucanía region. His poem-chant honors a family referent of solidarity and engagement with the struggle of the Mapuche across the lands and communities of Ngulu Mapu.

For readers familiar with Mapuche history, the anonymous "man" riding his horse recalls the Mapuche warrior of the sixteenth century commonly known in Spanish as Lautaro, a Hispanicization of Lautraro (or Leftraru), a key participant in the early years of struggle against Spanish colonialism.[7] As a young warrior leader (*toki*), he led the Mapuche in many victorious battles against the Spaniards that culminated in the execution of the conquistador Pedro de Valdivia in December 1553. Even though Lautaro was co-opted as an idealized epic figure in Ercilla y Zúñiga's *La Araucana* and later assimilated in patriotic Chilean discourses, Leonel Lienlaf brings the figure back into Mapuche community history in his poem by overlapping the historical figure of "Lautraro" and a family participant in late nineteenth-century Mapuche resistance against Chilean colonialism.[8]

In Lienlaf's poem, both figures return through motion—galloping, singing, and moving through writing into chant. In its considerable performative power, the motion of the "language of the land" or the "language of earth" (Mapudungun) rubs up against notions of textuality restricted to writing or to the format of the book. Ül—that is, the Mapuche chant—frames Lienlaf's poetry. Neither fully written nor fully oral, the poetic event is unsettled. It is a metonymy for the unsettledness of the Mapuche subject amid territorial crisis and struggle.

Expressed in the figure of "a man" galloping on his horse, the Mapuche being is performatively in motion "like the wind." This verse, "like the wind," entails materiality and ephemerality. The mobility of language and languages, the flow of images, and the passage from writing into singing join with the movement of the "man," the horse, the dogs, the wind, and Lautraro. The text allegorizes images of resurgence as well as mutation and transmutation in the uneven currents of history. The force of Lienlaf's poetry revives figures of a continuous history of

Mapuche anticolonial struggle by invoking Lienlaf's ancestor and Lautraro, the icon of the war against the Spanish invasion. These two figures of resistance emerge from the past and come to life poetically as part of a present in which new Mapuche struggles oppose corporate and state colonialism within a neoliberal Chile.

By staging Mapuche voices and sounds in the literary realm, Lienlaf's writing and oral performance made a remarkable contribution to the Mapuche sociopolitical movement just as it was acquiring visibility and audibility in the public arena in Chile in the late 1980s and early 1990s. Among Mapuche reader-listeners, his poetry awakens Susan Stewart's "aural connections." It also invites non-Mapuche audiences to cultivate the reading of poetry as a figurative mode of cross-cultural listening. There is a decolonizing force in the written and oral performance of this poetry. The wingka language is no longer the dominant code. Graphically, visually, and sonorously, Mapudungun is positioned as the primary language. In Lienlaf's verses, the dreams of the land (Mapu), its language of voices and images (dungun), and its people (che), creatively resound as writing and chant.

ÜÑÜMCHE, OR THE BIRD MAN IN LORENZO AILLAPAN CAYULEO'S POETRY

In December 2003, the Mapuche-Lafkenche author Lorenzo Aillapan Cayuleo published a collection of poems titled *Üñümche: Hombre pájaro* (Bird Man). An earlier version of the same work had been published in Havana in 1994 as *Hombre pájaro* and won the Prize for Literature in Indigenous Languages awarded by the cultural center Casa de las Américas in Cuba. Of the seventy-two poems that comprise the 2003 book, sixty-eight portray birds native to the river, lake, and coastal area where Aillapan Cayuleo was born and has resided most of his life; an area that is part of the Lafkenmapu (Land of the Coast), that is, a Mapuche Lafkenche territory (Aillapan Cayuleo 2017, 21–23).[9] As the title suggests, Aillapan Cayuleo not only approaches birds as a subject, in a referential way, but also establishes an embodied, mirroring relationship between the Native self (*che*) and the figure of birds (*üñum*). Indeed, for Aillapan Cayuleo, birds and nonhuman lives and languages are generally metonyms for the human. In this poetics of metonymical resonance, Aillapan Cayuleo centers the Mapudungun linguistic concept of *wünül*, which can be translated as onomatopoeia. In the prologue to his collection of poems, he highlights the effectiveness of this notion in acknowledging community affects and bonds:

> El sonido y el silencio alcanzan una significación singular, por lo que el canto *wünül*=onomatopeya y el hecho de reproducirlo, adquieren frente a la comunidad un lugar destacado, su entorno natural, la vida cotidiana, pena, alegría, amor al comienzo de la pareja humana y su prole. (Aillapan Cayuleo 2003, 10)
>
> Sound and silence achieve a particular signification; for this reason, the *wünül* chant=onomatopoeia and its reproduction acquire a remarkable status in the community, its natural environment, daily life, sadness, happiness, love at the beginning of the human couple and their offspring.

Through technical reliance on wünül, Aillapan Cayuleo engages with bird sounds in both his writings and public presentations in Mapudungun. In *Üñümche: Hombre pájaro*, the aesthetics of sound plays a critical role. Aillapan Cayuleo's poetry employs the Mapuche language as a space of onomatopoeic activism as well as of territorial engagement. In his study of the role of ideophones onomatopoeic utterances in Navajo poetry, anthropologist Anthony Webster points out that they mark "the sense of locality" that sound symbolism entails (2009, 77). In Aillapan Cayuleo's the linkage between bird singing and onomatopoeic sounds also vocalizes a particular sense of local territoriality in the Mapuche context. His poetics and phonetics as *üñümche* aurally embrace the singing of birds native to a specific territoriality, the Mapuche-Lafkenche lands and waters of Ngulu Mapu. At the same time, by onomatopoeically performing them his poetry positions local sonic ways (that of native bird singing) to withstand the ongoing acoustic and ecological destruction that has been taking place over the main lake in the area (Lake Budi) and its surroundings; an area that, at some point in history, had the serenity of vast, unperturbed waters.

The bird universe has been a central motif in poetry for Chilean Creole canonical authors such as Pablo Neruda. In *Arte de pájaros* (1966), however, birds remain objects under Neruda's gaze. For a less referential and more linguistically embodied poetic discourse, verses of Vicente Huidobro's *Altazor* (1931) come to mind. Following his avant-garde aesthetics of language, Huidobro approaches the movement and singing of birds playfully in "Canto IV." In his book *La nueva novela* (1977), the late twentieth-century avant-garde poet Juan Luis Martínez also speculates philosophically and poetically about birds as ciphered figures of chant and language. Closer to Aillapan Cayuleo's Indigenous poetics, K'iche' Maya author Humberto Ak'abal has explored the expressive possibilities of onomatopoeia to delve into the similarities between human and nonhuman, animal and nonanimal languages. His books *El animalero* ([1990] 2008) and *Hojas del árbol pajarero* (1995) include several poems in this vein.[10] In turn, however, Aillapan Cayuleo's *Üñümche: Hombre pájaro*, walks us through a distinctive Mapuche-Lafkenche route to speak, sing, and imagine *alongside* birds and the land.

Aillapan Cayuleo was born and raised in the region around Lake Budi, a tidal lake of brackish water near the Pacific coast in the province of Cautín. His community of origin and his current residence are near Puerto Saavedra, named after Colonel Cornelio Saavedra Rodríguez. This colonial marker reminds us that these lands, like most of the Mapuche territory, have been subject to continual settler occupation. The Mapuche-Lafkenche of the Lake Budi area experienced this colonial deployment through the intensification of land dispossession during the early decades of the twentieth century, when the Chilean state sponsored the establishment of colonizing agencies (*sociedades colonizadoras*) in Ngulu Mapu that administered a massive transfer of Native lands to Chilean and European settlers. The anthropologist Fabien LeBonniec has shown that in 1902 the Empresa Colonizadora del Budi (renamed the Sociedad Colonizadora del Budi in 1914) accelerated the invasive process of settler occupation and the subsequent alteration of the Mapuche lifeways in the Lake Budi region (2006, 555–58). Nonetheless, the area is still densely populated by Mapuche-Lafkenche communities, such as those where Aillapan Cayuleo grew up amid an extraordinarily diverse range of water and land-based bird life.

Aillapan Cayuleo's self-figuration as an *üñüm-che* characterizes his writings and public poetry readings. In *Üñümche: Hombre pájaro* he creates an aesthetics and a politics of resistance to the ongoing environmental and social destruction of the acoustic and physical ecology of Ngulu Mapu that resulted from multiple waves of capitalist modernization. For Aillapan Cayuleo, sound and the environment constitute an axis of Mapuche life and of Lake Budi life in general. His written and vocal eulogy of the sounds of birds features native species that are part of the living population around Lake Budi and other Mapuche lands and waters.

In Western literary theory, the lyric poet is situated at a distance from nature, which enables the expression of a human desire to reach toward the high spheres of music and the "divine." As described by American literary critic George Steiner, the stance elevates human beings over the "natural" and "animal" condition.[11] Cultivation of the music of language thus transports the poet to a dimension closer to the realm of the spirit than to matter. In Chilean poetry, *Altazor* exemplifies this elevation of the poet to the transcendence of the higher spheres. As a long lyrical poem, this canonical work from the avant-garde period in Chile tells the journey of Altazor as a modern Icarus who attempts to reach the sky and the sun—that is, the realm of the gods—before succumbing to a disastrous fall.[12]

Aillapan Cayuleo's poetry takes a radically different approach. Nonhuman and human expressiveness merge in a movement toward each other; they establish sensorial and territorialized relations. His land-based Mapuche-Lafkenche poetics diverges from the Western anthropocentric and teleological metaphysics of elevation that Steiner's theorizing represents. *Üñümche* implies an imagined yet corporeal bird-becoming of the human self, mirroring and resonating as an echo, and an *eco* (*oikos*: home), of the animality of languages always in touch with the body and matter, on the ground, over water, or in the air, while Aillapan Cayuleo's speaker positions himself as an observer of bird life on the land and the sea around the shores and the surface of Lake Budi's waters.

The relationship between birds and humans is central to Mapuche cultural life. *Choyke pürrün*, "the dance of the native ostrich," is part of a Mapuche ceremonial and celebratory event, namely, the Ngillatun, in which people circle a sacred tree, dancing while holding a light blanket over their backs in emulation of the wings of a bird in flight.[13] Drawing on his background as a native speaker of Mapudungun, Aillapan Cayuleo evokes this ceremony in "Tachi Choyke," followed by a version in Spanish titled "El avestruz" (The Ostrich):

> Ayekafe mülenke pu che rumel pürüfengelu
> ad entukey choyke tañi allangechi pürün
> feymeu feula fillantu reke mütrümütrümngekey
> pu liwen—rupan antü—ka pun llemay
> inautuñmakefiyiñ chumngechi tañi pürün

tachi alka ka tachi domo choyke engu fel. (Aillapan Cayuleo 2003, 100)

Para la danza de los pueblos originarios está presente
por su manera de andar el Avestruz es imitado en el baile
ahora todos los días se invoca, se acuerda
por la mañana, al atardecer y por la noche
con maestría imitamos la hermosa danza del ave
tanto el macho como la hembra frente a frente en el cortejo. (Aillapan Cayuleo 2003, 101)

It is present for the dance of original peoples,
for its way of walking the Ostrich is imitated in the dance
now every day is invoked, it is recalled
in the morning, in the afternoon and at night
we imitate the beautiful dance of the bird with mastery
both the male and the female face to face in the courtship.

Aillapan Cayuleo's poem reflects the symbolic and physical significance of the ostrich in Mapuche ceremonial culture. The poem extols the mimetic relations embodied in a Ngillatun between the figure of human bodies and that of the bird: "We imitate the beautiful dance of the bird with mastery / the male as well as the female facing each other in the courtship." The prominence of the birds in Mapuche imaginary is foregrounded during the Ngillatun ceremony, particularly through the dance of the ostrich. In the poem "Tachi Choyke," Aillapan Cayuleo praises the bird's walking and dancing as a sort of Mapuche mimesis. In *Üñümche: Hombre pájaro*, the identification of birds with human beings is also central to the sound sphere of language. Aillapan Cayuleo uses onomatopoeia as a graphemic and phonetic recourse, particularly in the texts in Mapudungun, such as the ending verses of the poem "Tachi Choyke" (The Ostrich),

Choyinkekey anay—choyinkekey anay
choyinkekey anay—choyinkekey anay. (Aillapan Cayuleo 2003, 100)

Every poem in Mapudungun incorporates onomatopoeia or similar techniques for suggesting sound. Aillapan Cayuleo's linguistic and phonetic skill brings the written text close to bird singing. Strikingly, the versions in Spanish do not use onomatopoeia but rather depict some aspects of the poeticized birds in language that is more literal than figurative, more descriptive than performative. Employing different inflections across languages, poetry as phonetics takes full shape in the Mapudungun texts, giving the Indigenous language a distinctive force in rhythm and sound. The expressive difference from Spanish underscores how linguistic indigeneity becomes poetically and sonically invigorated.

Aillapan Cayuleo's phonetic expressiveness gains more resonance in declamations in which the poet stretches the phonetic possibilities of his voice to mimic the sounds of birds—from whistling to guttural grunts to soft whispering. In Aillapan Cayuleo's live presentations, onomatopoeic turns become playfully transported into the domain of voice, as registered in *Wünül: Concierto de pájaros* (Onomatopoeia: Concert of birds), a 2008 documentary about his verbal, vocal, and physical engagement with the Lafkenche territory soundscape of Lake Budi. This thirty-three-minute video, produced by Delestero Realizadora, a local independent studio, introduces us to the persona of Lorenzo Aillapan Cayuleo as a bird man (üñümche) and his relationship to the oceanic and coastal environment of Lake Budi, Füdi Leüfü in its Mapuche denomination. The video opens with the roaring of the Pacific Ocean and views of its silver-colored waters into which the lake flows. Aillapan Cayuleo walks us through the natural spots of Lake Budi, following the trail of the native birds' chants and sounds while speaking about the significance of bird sounds as integral to the acoustic and semantic ecology of Lafken-Mapu, the coastal Mapuche territory. Aillapan Cayuleo states, "Ningún pájaro canta por cantar sino que entregan un mensaje onomatopéyico" (Birds don't sing just to sing; they deliver an onomatopoeic message).

Aillapan Cayuleo's narrative introduces viewers to his poetics on the Lake Budi's sonic environment. *Wüñül: Concierto de pájaros* concludes with his ül (chant) titled "Amulepe." In its main passages, the lyrics encourage audience participation:

Amulepe, amulepe ülkantun,
amulepe kojagtun,
amulepe pentukun,
amulepe nütramkan.

> Let's continue the chant,
> let's continue the discourse,
> let's continue the greetings and introductions,
> let's continue the conversation.

The final verses reiterate the invitation: "Amulepe ülkantun, / amulepe ayekan" (Let's continue the chant, / let's continue). All this seems to be a narrative and a performance by Aillapan Cayuleo problematically marked by a sort of comic self-exoticization of the relationship between the Mapuche-Lafkenche and the natural environment. This recapitulates the public perception and interpretation that has prevailed over Aillapan Cayuleo's self-figuration as a "bird man." Yet, as his audiovisual narrative evolves, a broader historical resonance of his poetics and politics becomes evident. By the end of the video, Aillapan Cayuleo recalls the Indigenous identity of Lake Budi as well as the waves of colonial settlers who historically have besieged it.

> Füdi Leüfü es su nombre. . . . Otras gentes, otras gentes que han venido pero nunca han querido respetar . . . te han destruido. ¡Se perdió el Ruleketrawe! Se perdió la vega donde hacíamos la segunda cosecha. ¿Por qué ha perdido? A lo mejor por la conducta de los seres ambiciosos . . . que quisieron adueñarse de este lago, llenando de motonave, con barcaza, con puentes exóticos, con luces, que no es nuestro . . . (Aillapan Cayuleo 2008)

> Füdi Leüfü is its name. . . . Other people, other people have come here . . . but they have not respected you . . . they have destroyed you. The Ruleketrawe is gone! . . . The wetland where we carried out the second harvest is gone. Why has the lake lost? Maybe it's because of the behavior of ambitious beings . . . who wanted to take possession of this lake, populating it with motorboats, with barges, with exotic bridges, with lights, which are not ours . . .

Reading and listening to Aillapan Cayuleo's poetry invites scholarly echoes on the homologies between the language of birds and human beings. Steven Feld's classic *Sound and Sentiment* (1982) offers an ethnographic journey into the life of the Indigenous Kaluli community in New Guinea. According to his study, the sentimental expressive system of birds gives shape to a cross-fertilization between the linguistic and communicative realms of the human and bird communities. Aillapan Cayuleo's poetics on the relationship between native birds and the Mapuche parallels the conditions Feld described among the Kaluli, except that Mapuche politics of the land strongly resonate in Aillapan Cayuleo's book, whereas any Indigenous politics is absent from Feld's analysis. Indeed, as a poet situated in his local environment, Aillapan Cayuleo lays claim to an affective and territorialized relationship between birds, the community, and his persona, which echoes the critical place in the contemporary Mapuche movement of the concept of territory (the Mapu) as a human and nonhuman living environment. In this sense, his poetics of bird singing is political and cannot be reduced to an exclusively "sentimental" and "cultural" linkage to the local soundscape.

EARTH RESONANCES, ONOMATOPOEIC ACTIVISM

Since the 1980s, Mapuche organizations and leaders have centered their concept of territory in their political discourse. Territory here has various dimensions for the Mapuche, ranging from the spiritual, linguistic, and cultural to the social and political. By the 1990s, "the reconstruction of territorial identities" had gained prominence as an idea and a historical endeavor across communities, "starting with repositioning traditional leaderships, generating a process of cultural reconstruction, reengaging their ceremonies with a much more original sense, as a public space in which the Mapuche express their religiosity and taking up again their social learnings as a People" (Mariman Quemenado et al. 2006, 268). This process gained traction among the Lafkenche, the coastal branch of the Mapuche people to whom Aillapan Cayuleo and his Lake Budi community belong.

According to Chilean administrative geography, the Lake Budi area is part of the Province of Cautín. Even within wingka language, as part of regional toponomy, the denomination Cautín survives as a linguistic trace of the significance of bird life in the area surrounding the lakes and seashores where the Lafkenche people live. Indeed, it is a Mapudungun word that refers to a type of wild duck that has long inhabited the rivers, lakes, and other water lands of the region. It is also the name of one of the main rivers in the province, the Cautín River. The poem "Kautin/Wakeñ" honors the territorial presence of this native bird Aillapan Cayuleo translates *kautin* or *kautiñ* into Spanish as *pato overo* (dappled duck). His verses exalt the duck's omnipresence in the work of local clay potters:

> Fayti pu wüdüfe adtukuy metawe meu
> tachi kautiñ wakeñ leufümangiñko üñüm may
> feychiwüla küllfü pingekey feyti metawe . . . (Aillapan Cayuleo 2003, 112)

> El curioso alfarero lo modeló para cántaro de greda
> cuya figura de ave quedó inmortalizada hasta hoy
> el cántaro llamado "Küllfü" es el canto onomatopeya
> de esta ave . . . (Aillapan Cayuleo 2003, 113)

> The curious potter modeled it for a clay pitcher
> whose bird figure was made immortal until today
> the pitcher called "Küllfü" is the onomatopoeic chant
> of this bird . . .

These verses refer to the bird-shaped pitchers traditionally made by local potters. Clay or wooden figures of the *kautiñ* or *kautin* are a common souvenir in marketplaces in southern Chile, especially in cities like Temuco and Valdivia. Aillapan Cayuleo praises the omnipresence of the kautiñ in local pottery as an inspiring motif and as the bearer of a distinct chant, namely, *Küllfü*. With its sound in mind, Aillapan Cayuleo's poem in Mapudungun includes a refrain that is repeated three times before the poem concludes:

> Küllfü kur kur—kautiñ kautiñ kantiñ
> Küllfü kur kur—kautiñ kantiñ kautiñ. (Aillapan Cayuleo 2003, 112)

In these verses, *kautiñ* is written twice as "kantiñ," which phonetically makes it sound like *kantun* (singing, in Mapundungun), and like *canto* (chant, in Spanish). "Kautin/Wakeñ" and the poems of *Üñümche: Hombre pájaro* constitute, about all, a sonic mapping of the bird environment of the Lake Budi area. Thus, like a *kautiñ*, the poetic persona of Lorenzo Aillapan Cayuleo moves through its Lafkenche territoriality; and, in this motion, he mutates into multiple other bird forms. The collection of poems begins as a eulogy to the native bird *tremka*. Poet and bird replicate each other, as manifested in the very title of this opening poem: "Tachi tremka / tremkakawün"—in Spanish, "La tenca (pájaro poeta)" (The tenca [poet bird]). The verses confirm this homology.

Doy fel tañi küme ülkantun
tachi üñüm koyautukelu / ülkantukey / fill trokiñ tripantü
yafüngen / leuliñkulen / kuralen / ka chigagken
pañilwelen / wünelen / kachigagken
. . . (Aillapan Cayuleo 2003, 12)

Que sea muy ansiada la canción
del pájaro poeta que canta por todas las temporadas
soy duro e invariable, compacto y sonoro
soy resistente, soy principal, y más que poderoso
. . . (Aillapan Cayuleo 2003, 13)

That it be very desired the bird poet's song
who sings across all seasons
I am tenacious and persistent, compact and sonorous
I am resilient, I am a headman, and more than powerful . . .

The trope of "poet-bird," analogous to "bird-man," is affirmed in the quotation. The first two stanzas are impersonal exaltations of a "bird poet who sings across all seasons." The human and nonhuman animal selves are melded through the common virtue of being able to generate a chant and to practice chanting continuously. More acutely, the very grammar of the poem changes. The speaker leaves behind the impersonal posture and embraces the first person: "I am tenacious and persistent, compact and sonorous / I am resilient, I am a headman, and more than powerful." Multiple virtues, such as endurance and persistence (existential and ontological), "sonorous" (aesthetic) and "powerful" (social) are foregrounded by Aillapan Cayuleo's "bird poet" figure. In the remaining stanzas, this poetic human/nonhuman literary persona describes its ethics of community and land relations,

Tralka matra Kinturay / allangechi trüfuyal
ka fill rakiduam ülkantuy: Wechulen lof Mapu meu
tachi pu che küdaukelu rumel / walüngmalekey

pütrü müna weraley alka trüri
tralkan reke kümülli ñi miyauwün
kañpüle Mapu tuwünche fapüle turpe ñidolkaupalayay.
Kotrotron / Trayay / Trililin Newengen / Yafüngen / wünelen
Deumakan / Sebastian / fütra trontron / kinturay Tralka
matra / fütra
dollüm. (Aillapan Cayuleo 2003, 12)

De canilla como escopeta Kinturay, divinamente acariciable
y canta pensamientos: Crecer, crecer en la Comunidad
el que trabaja la Tierra estará repleto de alimentos
valiente como un gallo
camina retumbante como un trueno
gentes de otras comarcas jamás se adueñarán de este suelo. (Aillapan Cayuleo 2003, 13)

Of Kinturay shotgun-like shinbone, divinely lovely
and it sings thoughts: To grow, to grow up in the Community
the one who works the Land will be full of nourishment
brave like a rooster
he walks resounding like a thunderstorm
people of other villages never will be able to take over this soil.

In these lines Aillapan Cayuleo weaves together three concepts that are crucial to Mapuche lifeways: *Rakiduam* (thought), Lof Mapu (territory-community), and Mapu (land, territory, universe). Within this conceptually dense triangulation, the bird poet's ability to deliver "thoughts" through song (*rakiduam ülkantuy*) is first praised. After the colon, Aillapan Cayuleo expounds on the thoughts: the need for any Mapuche person to grow, work, and be rooted in a community (Lof) and its land (Mapu). As a sort of fable in verse, the poem presents a moral lesson. The person "who works the Land" will become a "brave," "resounding" presence. Cultivating land relations leads to a cross-fertilization between the Mapu and the Che to which communal well-being and self-care are intrinsic. The speaker's territorial engagement generates an anticolonial shield for the self and the land, a force that is physical, psychological, and acoustic. It will prevent potential settlers from appropriating Native "soil." The strength of the Mapuche self takes on a sonic shape, a way of walking that resounds "like a thunderstorm": "tralkan reke kümülli ñi miyauwün."

Resonance is not foreign to Mapuche modes of apprehending and connecting to earth. In a key moment of the Ngillatun, that important Wallmapu ceremony, community members ride several horses to gallop around the ceremonial site.[14] As the movement of horses grows more forceful, the attendees can hear the rumble from underground. The event features the omnipresence of the land as a living sonorous territory. An analogous logic of resonance defines the nature of *kultrun*, the circular drum used in Mapuche religious, social, and even political gatherings. For us, the Mapuche, kultrun is not merely a "musical instrument"; it has far deeper significance. Indeed, the Four Directions of Earth (Meli Witran Mapu) are drawn on the kultrun's surface—so, figuratively, this is what sounds. As Aillapan Cayuleo states, the sound of kultrun "is the sound of earth." He points to the onomatopoeic linkage between kultrun and the land: "Earth has a musicality *kultxug, kultxug, txulum, txulum, kultxug,* always changing" (Aillapan Cayuleo 2017, 178).

The accumulated sonorousness of these lines leads Aillapan Cayuleo's poem in Mapudungun to end in verses with an onomatopoeic, sonic force. As in the sound of the word *kultrun*, the phoneme *tr* (or *tx*, in other alphabets), so peculiar to Mapudungun, stands out in the musicality of these final verses: "Kotrotron / Trayay / Trililin Newengen / Yafüngen / wünelen // Deumakan / Sebastian / fütra trontron / kinturay Tralka matra / fütra // dollüm." These utterances function as fragmentary verbal expressions, resounding on the page like a thunderstorm or the beats of the kultrun, the sound of earth. Translated into Spanish, the poem's effect disappears, and the text is shorter.

Mapudungun's resonant logic of land-language relations fuses voice and sound as a composite term (*Mapu*: land, earth, territory, universe; and *dungun*: language, voice, sound, sense). These relations include the language and sound of the whole Mapu. As in Lienlaf's poetry, these relations are consequential elements of Aillapan Cayuleo's writings and presentations. Again, like Lienlaf, the prominence of the human being–bird homology as part of a territorial and environmental resonance gains full force in Mapudungun, conveying how the language defines voice, sound, and sense as linked to place and space. The language emerges from the bird-singing-ecology of the territory from which the poet writes and speaks, that is, Lafkenmapu—the lands and waters of the Mapuche-Lafkenche. Through the graphemic and phonetic turns of language in more than seventy poems of Aillapan Cayuleo's book, this acoustic ecology resonates much more in Mapudungun than in Spanish, where it manifests markedly different expressive qualities. The colonial language is not merely displaced and accorded supplementary status. It fulfills a descriptive and literal function that lacks major rhythmic and phonesthetic elaboration, while the language of the colonized vibrates and reverberates poetically.

The poetics of the üñümche is crafted from a cross-fertilization between the linguistic, musical, and aesthetic possibilities of Mapudungun. Unlike monolingual discourse, its written and vocal practice is dislocated between languages. The dissonance in the gaps between the languages and the onomatopoeic turns and final verses in Mapudungun illuminates the boundaries between the "acoustic territories" of Lafkenmapu and those of the colonizing society.

According to Aillapan Cayuleo, people who come to his bird man presentations tend to caricature them as a joke. Or, somewhat more generously, they consider bird singing the comic side of an exotic cultural difference within the taxonomy of the settler colonial "intercultural" museum of neoliberal Chile. Aillapan Cayuleo's vocalization of bird singing is playful and at times humorous, which can lead some audiences to laugh and ridicule or perceive the presentation as an exotic spectacle of the "hombre pájaro." After one such performance at a community gathering in Temuco, Aillapan Cayuleo told me, "Some people only laugh at it because they don't understand it." Laughing, joking around, is, in fact, part of Aillapan Cayuleo's playful art. Yet there are also subtle dimensions to his work that less critical audiences overlook.

In a three-stanza poem dedicated to the black-neck swan of the region, titled "Tachi Piupiukürüpel—Piupiukürüpelkawün" / "Cisne de cuello negro," the last stanza poignantly recalls the settler colonial history that has disrupted the lives of the Native human and nonhuman lifeways in the Lake Budi area. Here I quote the relevant stanza from the Spanish version:

> Ya hace más de cien años llegaron los afuerinos "wiñka"
> a este paraíso Mapuche y trajeron "una desgracia"
> con tantos botes y pescadores alterando a las gentes del lago
> ellos corretearon a esta gran ave con sus ires y venires
> a la gallina acuática Mapuche, cisne de cuello negro
> por eso es lastimero su canto de ida y vuelta. (Aillapan Cayuleo
> 2003, 33)

It has been more than one hundred years since the "wiñka"
outsiders arrived
in this Mapuche paradise and they brought "a tragedy"
with so many boats and fishermen disturbing the people of
the lake
with their comings and goings they chased this great bird
the Mapuche aquatic hen, the black-neck swan
therefore its back-and-forth chant is pitiful.

The first lines recall the era of the mid-nineteenth-century Campaign of Pacification of the Araucanía and the Chilean state-sponsored arrival and establishment of White Creoles and new European settlers ("the 'wiñka' outsiders") in the Mapuche territory. The colonial occupation deployed military force throughout Ngulu Mapu. The invasion drastically altered what Aillapan Cayuleo characterizes as a "Mapuche paradise," particularly in reference to Indigenous patterns of life on and around Lake Budi, and it changed Mapuche history. "With so many boats and fishermen disturbing the people of the lake / with their comings and goings they chased away this great bird / the Mapuche aquatic hen, the black-neck swan / therefore its back-and-forth chant is painful." The metonymy, equating the fate of the people (Che) and that of the "aquatic" bird, allows Aillapan Cayuleo to portray the colonial situation as disruptive to living conditions for both "the people of the lake" (Lafkenche) and the black-neck swan. The "pitiful" tone fissures the very "chant" of the black-neck swan, imparting a harrowing resonance that the Spanish version alludes to but that, in Mapudungun, becomes more dramatically embodied performatively through the onomatopoeic turns:

piu piu piu piu wikür wikür wikür wikür
piu piu piu piu wikür wikür wikür wikür. (Aillapan Cayuleo
2003, 32)

A traumatized chant of the black-neck swan emerges from Lake Budi, in resonance with colonial history. In these lines, the wobble in the vowel sound (*iu; i-ü*) heightens the wounded tone, the sensation and echoing sound of the tear. The "pitiful" chant of the black-neck swan emerges as a dissonance in the "concert of birds" that Aillapan Cayuleo's poetry registers, while both the acoustic ecology of Lafkenmapu and its language have been altered undeniably.

The poetry of Aillapan Cayuleo and Lienlaf brings into play Mapuche sound culture as an environment of human and nonhuman, tangible and nontangible relations. Both poets communicate an acoustic ecology in which animal, human, physical, and spiritual planes of sonic activity intersect and engage with one another. Chilean scholar José Pérez de Arce states, "Birds, animals, rivers and winds, besides sounding like what we hear, resonate as the chants of the spirits that animate and protect them, and these chants are imitated by the human" ([2007] 2020, 71). In *Pewma dungu / Palabras soñadas*, Lienlaf alternately incorporates, invokes, and exalts a universe that celebrates varied relationships between the poetic first person, singing, the human and nonhuman animal community, and the natural environment. Lienlaf thus makes us participants in a time and space of sensory and symbolic conjunction between the Mapu (land/territory) and dungun (voice/language). At the same time, many of these poems read as an allegory of the invasive arrival and impact in Ngulu Mapu of the forestry industry, a corporate force that is part of a Chilean recolonization of Mapuche lands under the prevalent depredatory and extractivist neoliberal capitalism of late twentieth- and early twenty-first-century Chile.

Amid these disjunctions, Lienlaf's poetry passes through a heterogeneous universe of sounds and voices. This is what is expressively concentrated in its title *Pewma dungu / Palabras soñadas* (Dreamed Words). In this new corpus, many of the singing poems are inscribed on the page as music, sound, and ceremony. Each of Lienlaf's poem-songs is revealed through chants and dreams as a form of trance.

Konan tachi ülkantun pewmamu,
düngunpewmamu
chonkitunmew kachill kütral
Wekun ta kürrüf
pürulmekefi mawida. (Lienlaf 2003, 6)

Me adentro
en estos cantos de sueños,
dormitando cerca del fuego
mientras afuera
el viento
hace bailar a las montañas. (Lienlaf 2003, 7)

I go into
these songs of dreams,
sleeping close to the fire
meanwhile, outside
the wind
makes the mountains dance.

Song ("into") and dance ("outside") form part of a simultaneous movement that couples the speaker's introspective journey—into "songs of dreams"—with a physical environment that takes the shape of being-in-motion. Lienlaf presents the realm of pu pewma (dreams) as an inner space in which his chants take place. Mapuche pedagogue Ramón Curivil Paillavil explains that, for the Mapuche, dreams are "an instance of privileged knowledge" where "a communication between the world of here and the other world, between the present and the future" is established (2002, 34–36). Lienlaf defines his own poetry as a "song of dreams" that transports the self to another dimension yet remains metonymically contiguous with the music (of "the wind": *kürrüf*) and the motion (of "the mountains": *mawida*) of the tangible, material environment. Through the coupling force of the word "meanwhile," the text remains in step with the movements of an interior-exterior environment driven by the rhythm of pu pewma. In Western lyrics, the trope of the dream is associated with the romantic notion of inspiration or numen, as a daydream in the upper spheres of Parnassus. In Lienlaf's poetry, the trance of the poem occurs in the sensorial, corporeal, and earthly register of the Mapu. His poetic journey travels through a heterogeneous territory of voices, sounds, and images attached to the land, where whispers, screams, creaks, applause, sobs, groans, silences, and noises happen and flow from the earth-tongue. This is what can be listened to and heard throughout the corpus of *Pewma dungu / Palabras soñadas.*

However, toward the end of the volume, real-world happenings intrude. Images of economic modernization break the tone, attitude, and lyrical register of the speaker. The Mapuche chant—along with its dreaming subject—collapses as the last set of poems describe territorial fracture and crisis, as evidenced in the poem "Wedake dungu" / "Imágenes" (Images):

Pen wedake pinu
purumeken wente ngenochi ko

Pen
chumechi
ñi nümameken kay
inche trufken. (Lienlaf 2003, 46)

Veo
ejércitos de pinos
bailando sobre los restos del estero
y camiones blindados
empolvando las espaldas
de kai-kai. (Lienlaf 2003, 47)

I see
armies of pine trees
dancing on the remains of the stream
and armored trucks
dusting the backs
of the kai-kai serpent.

Rather than a litany of sounds, the poem becomes image. The title, "Images," makes this transition literal. The Mapudungun verb *pen* (see) stresses the visual turn, as if it constituted the "chronicle of the seen and lived," in the manner of a colonial chronicle. Such shifts are common in Lienlaf. His first collection of poems included drawings, and Lienlaf's poetry often juxtaposes oral and sound imaginaries with visualization of the territory. "Images" let us "see" the invasive presence of the forestry industry in the Mapuche environment: "Veo / ejércitos de pinos / bailando sobre los restos del estero / y camiones blindados." These lines evoke the specter of military force historically deployed by the authoritarian and repressive regime led by General Pinochet to impose the neoliberal economic model and the forestry industry on Ngulu Mapu / Chile, as well as the mobilization of militarized police in the post-dictatorship era over those Mapuche communities engaged in land-back struggles. The figures of "armies" and "armored trucks" are synecdoches of these settings of state violence. In a parallel image, the "pines" function as a more literal reference to the forestry industry. Even in the more democratic climate of the early 2000s, these images depict the ties between the national and transnational forestry corporations and recall the lasting neoliberal legacy of Pinochet's military regime.

Since the neocapitalist transformations in Chile of the 1970s, noise and pollution from the forestry companies' machines have been significant factors in the Mapuche region. The Pinochet regime's Decree 701 enhanced the growth of forest pine and eucalyptus plantations at the expense of native forests by exempting forestry companies from taxes. This had a major impact on Mapuche territory. By the mid-2000s, corporate forest monoculture exceeded four million acres between Regions VII and X in southern Chile. The forestry industry's invasion began in the Bío-Bío River area in the north, where the Arauco region begins, and then spread through the provinces of Concepción and Temuco, reaching into the provinces of Valdivia and Osorno farther south—in other words, into the historical Mapuche territories.

In the last lines of "Wedake dungu" / "Imágenes," Lienlaf emphasizes the pollution from industrial forestation. "Armies" and "armored trucks" end up "powdering the backs / of the kai-kai serpent." In Mapuche culture, kai-kai is a serpent figure that rules the waves of the rivers and the sea, a water snake that is the counterpart of the treng treng serpent that takes the shape of hills in the Lafkenche territory. In Lafkenmapu, the coexistence of both serpents nurtures environmental balance and is responsible for the *küme mongen*, that is, what the Mapuche value as living well. Corporate pine and eucalyptus plantations have deteriorated aquifers and groundwater flows, an ecological and human disaster that affects local mestizo and Indigenous communities. Lienlaf's poetry revolves around images of environmental crisis: "the remains of the stream" and the dusty "back of the kai-kai serpent." Toward the end of *Pewma dungu / Palabras soñadas*, the noise of forest industries that invade Native territory in present-day Chile crack the Mapuche chant once again.

"Quinquen," one of the final poems in the book, records this crisis and fear. Quinquen alludes to the hiding and self-preservation that was part of the Mapuche experience during the genocidal Chilean invasion of Ngulu Mapu between the early 1860s and the 1880s. With its military, economic, and religious agents, the colonial occupation of the Mapuche Country was also the beginning of human, animal, and environmental devastation. The environmental history continues into the present neoliberal era. Haunted by that history, the speaker of the poem "Quinquen" trembles in fear. The sound of water emerges as a last refuge in Lienlaf's verses:

Quinquen Mapupüle
ñi aiwiñ
rofülerpuy ñi llükan. (Lienlaf 2003, 52)

Quinquen
y el miedo de mi sombra se durmió
abrazado por el canto del estero. (Lienlaf 2003, 53)

Quinquen
and the fear of my shadow fell asleep
embraced by the chant of the stream.

These verses illustrate the fear that drives the speaker to find refuge in the affective and sonic space of the "chant of the stream." Supported by the fluidity of the oral tradition, the poet calls on this liquid space to emphasize inherent links between self, language, nature, and territory through chanting and staging a constitutive bond of language (dungun) and the environment (Mapu).

FINAL REVERBERATIONS

A linguistic and territorial engagement runs through the poetically resonant writings of Leonel Lienlaf and Lorenzo Aillapan Cayuleo. Whether from the Alepue community (Lienlaf) or from the Lake Budi area (Aillapan Cayuleo), poetic language is permeated by a polyphony of voices that emerge from the waters and lands of the coastal territories of the Mapuche-Lafkenche, that is, Lafkenmapu. In coupling close reading with "attentive listening" (allkütun), my analysis highlights these dimensions of territorial resounding in two extraordinary exponents of contemporary Mapuche poetry.

The writings of Mapuche poets such as Lienlaf and Aillapan Cayuleo have often been co-opted by critics and readers whose romanticizing and exoticizing approaches place their works in a symbolic and cultural realm that is disconnected from the colonial histories of what is today called "Chile." My practice of allkütun enables me to pay close attention to the subtleties with which poetic language confronts and interrogates historical power relations in Ngulu Mapu. Allkütun invites readers to listen to the broader resonances of the environment. The male self that dominated the Mapuche literary scene of the 1980s still marks the poems' actual and symbolic positioning. Even so, by establishing a poetics of listening to the land (Mapu) and the people (Che), these poets pave the way for the polyphonic set of voices and bodies of the Mapuche literary and artistic movement that bloomed in the late twentieth and early twenty-first centuries.

The poetry of Leonel Lienlaf and Lorenzo Aillapan Cayuleo are part of a broader current of Mapuche sonorous agency on the mediascape. Rooted as they are in the ethos of Mapudungun, their literary writings and vocal arts express a relationship between language, sound, and the environment that is at the heart of the Mapuche ethos. Cultivating Mapudungun's territorial resounding enables these poets to strengthen their land-based poetics. Lienlaf's and Aillapan Cayuleo's writing and public presentations incorporate Spanish as a domain of semantic and phonetic resonance. Their poetic voices interfere with the scripts of linguistic and acoustic colonialism in Chile, capturing a complex overlapping of languages. Oscillating between chants, tears, and dreams, their writings join a Mapuche discourse fractured by the noises of colonial history.

....

For so many Mapuche Willliche and Lafkenche who grew up as I did in an atmosphere of Catholic evangelization and state-based Chileanization, in which practice of the native language, traditional ceremonies, and self-governance structures were restricted, the poetic writings of Leonel Lienlaf and Lorenzo Aillapan Cayuleo were much more than graphemic events on the page. Their voices, chants, or onomatopoeic vocalizations began to resonate in the late 1980s as sounds of a literary and cultural uprising that opposed the acoustics of Chilean colonialism. They challenged settler domination and the internalized linguistic and cultural colonialism among Mapuche who did not grow up speaking and singing in a Mapudungun environment. Such was my own experience, as someone who was from a Mapuche-Williche reynma *(kinship) that became part of a* reducción *(a shrunken or reduced territory). After many years of Chilean schooling and then moving into US academia, it was paradoxically the world of the* lifru *(book) that helped "the bird of my heart" take flight and reconnect to being Mapuche.*

By the beginning of 1998, while I was taking the initial steps to learn Mapudungun, I decided to write a paper on De sueños y contrasueños azules *(1994) by Elicura Chihuailaf. This got me into closely reading and listening to the Mapudungun versions of his poems. In the early 1990s, I had already heard Chihuailaf recite them in Mapudungun at several public presentations. I had also started paying attention to the distinct features of sonority that surfaced in the writings and voices of Lienlaf and Aillapan Cayuleo. But then, in July 1998, a key event occurred on this journey, when I presented on the poetry of Chihuailaf on a panel at a literature conference at the Catholic University of Chile, in Santiago. Immediately after the panel, a* peñi *(Mapuche brother) approached me to extend an invitation to discuss Mapuche literature on a radio show that broadcast on Saturday nights over Radio Yungay. He was peñi Ramón Curivil Paillavil, a founding member of the Mapuche radio program* Wixage Anai, *which, like the poetry of Lienlaf, Aillapan Cayuleo, or Chihuailaf, also used Mapudungun and Spanish. When I entered the studios of Radio Yungay that winter of 1998 in Chile, the soundscape of Wallmapu expanded for me. In my ears, the sounds of Mapudungun extended from literary form to the radiophonic ether.*

FOUR

....

Wixage Anai: Mapuche Voices on the Air

It is 10:00 p.m. on Wednesday, June 26, 1993. Ramón Curivil Paillavil, Clorinda Antinao Varas, and José Alfredo Paillal Huechuqueo, a team of self-taught Mapuche broadcasters, sit expectantly around the main table at the studio of Radio Nacional de Chile, on Lastarria Street in downtown Santiago. The first broadcast of the radio program Wixage Anai *is about to start. It is a cold winter night at the time of year when the Mapuche people celebrate the start of a new cycle of time that the Mapuche call Wiñol Tripantü, or We Tripantü. With Radio Nacional's wide reach through long and shortwave frequencies, the Mapuche radio program will likely be heard by audiences throughout Ngulu Mapu (Lands of the West, or what today is called central and southern Chile) and Puel Mapu (Lands of the East, or what today is identified as southern Argentina), which together constitute Wallmapu—the Mapuche Country. At this time of night, radios are on in many homes and workplaces, including those of Mapuche families and individuals. First, they hear a kull kull—a wind instrument that calls the community to gather. Then the drumming of a kultrun comes on. It is a Mapuche musical curtain opening. The sounds of the kull kull and kultrun are followed by the agglomerated and vibrant voices of an* afafan, *the collective way to motivate someone in Mapudungun:* Ya-ya-ya-ya! *Next, the voice of a Chilean broadcaster announces, in Spanish: "Radio Nacional de Chile presenta* Wixage Anai: Álzate y despierta." *Finally, the voice of Clorinda Antinao Varas comes across the airwaves: "Mapuzugun mew kimaiñ ta iñ tuwvn, amulzuguleaiñ pu lamgen, kimaiñ kvmeke zugu, kvmeke rakizuam, epvñpvle ta kimvwaiñ fei mew ta ayigeai ta Mapuce mogen. Wixage Anai!" The Chilean broadcaster reiterates the same message in Spanish (here in translation):* "Wixage Anai *[Get Up, Wake Up!] is a Mapuche program to learn about our roots, to foster our culture, our philosophy; to strengthen the bonds of unity of our People and to animate Mapuche lifeways in the city and in the country." This is the sonic and vocal opening of* Wixage Anai. *At that point, the Mapuche voices of Curivil Paillavil and Antinao Varas take over the broadcast. The radio show begins.*

....

As a program produced, managed, and broadcast by a team of Mapuche activists, spoken in Mapudungun with Spanish translation, *Wixage Anai* was the first radio show of its kind aired on a radio station with national and transnational reach. With the "Get Up, Wake Up!" slogan, the team signaled its intention to practice Indigenous radiophonic activism, mixing Indigenous Mapuche chants and music with news on the contemporary territorial, political, and cultural struggles of urban and rural life. *Wixage Anai* was shaped by the historical context of the 1990s, in which the flags of "self-determination" and "autonomy" dominated the Mapuche movement.[1] Thus, *Wixage Anai* was invested in making radio with a Mapuche politics and aesthetics of communication.

Wixage Anai articulated the sense of awakening that drove Mapuche political discourse in the late 1980s and early 1990s. For one of its members, Margarita Elizabeth Huenchual Millaqueo, the program was most remarkable for how it constituted a medium of "cultural recovery." As she explained to me, this radio show "was part of the Mapuche resistance. Ultimately, it was a program where the Mapuche language continued being spoken; an independent program that was free to address themes relevant to the Mapuche world" (Huenchual Millaqueo, interview by author, 2018). Huenchual Millaqueo highlighted how the Indigenous use of radio paralleled the politics of autonomy the Mapuche embraced.

From its first broadcast in 1993 to its last in 2019, *Wixage Anai's* team of producers and broadcasters consisted of Santiago-based Mapuche educators and community activists. They included founding members Ramón Francisco Curivil Paillavil, Clorinda Antinao Varas (referred to as Clara Antinao Varas from here on), and José Paillal Huechuqueo, during the early years; then, during the 1990s, María Catrileo, Margarita Elizabeth Huenchual Millaqueo, Elías Paillan Coñoepan, and Ricardo Tapia Huenulef; and, finally, Richard Curinao and Javier Salazar Cuminao, who joined the radio show in the late 2010s.[2] The group also established the Jvfken Mapu Center of Mapuche Communications as a legal entity to address administrative matters and fundraising as well as to conduct research projects critical to the Mapuche community in Santiago.[3] Through the Jvfken Mapu Center, the radio team also benefited from the additional support of other members of the Mapuche community in Santiago.[4]

The *Wixage Anai* radio collective drew on earlier Mapuche experiences in radio. Mapuche writers, educators, and leaders first ventured into this medium in 1990, via the Ñielol Radio Station in the city of Temuco, the urban epicenter of Mapuche life in Ngulu Mapu. There, on March 24, a collective led by Armando Marileo and Francisco Kakilpan initiated a ten-minute radio show embedded in a longer program named *Más que música* (More than music). As the Mapuche segment expanded, its prerecorded radio shows were also broadcast on stations in the provinces of Cautín and Valdivia, in Ngulu Mapu. Still, the very idea of the Mapuche using radio autonomously was inconceivable for radio station owners in Chile. According to Kakilpan, this early pioneering experience helped legitimize the use of radio for the emerging Mapuche movement by establishing "the idea of a Mapuche radio station" as something Wallmapu had needed for a long time (interview by author, 2010). The experience was a reference for the *Wixage Anai* team, who consulted with Marileo. Another influence on *Wixage Anai* was the late 1980s radio show that Curivil Paillavil helped lead for a couple of years at the same station in Temuco. The program addressed the *comunidad campesina* (non-Indigenous and Indigenous rural community) but nonetheless had significant Mapuche content (Curivil Paillavil, interview by author, 2022). In Santiago, Curivil Paillavil also initiated an early-morning radio show at Radio Colo Colo in 1993, a station with nationwide reach. Titled *Amulzuguleaiñ* and addressed to a Mapuche audience, this show was mostly in Spanish and lasted only a year. Nevertheless, this initiative at Radio Colo Colo gave Curivil Paillavil important media experience in Santiago as well as the opportunity to collaborate with Elías Paillan Coñoepan, who had studied journalism and radio communications and later became a member of *Wixage Anai*. Spurred by these earlier or parallel ventures, Curivil Paillavil led the idea to create and launch *Wixage Anai* as a Mapuche radio program conducted in both Mapudungun and Spanish, broadcasted from Santiago, and projected to audiences beyond the settler-state borders of Chile and Argentina into the acoustic territories of Ngulu Mapu and Puel Mapu—that is, to Wallmapu as a whole.

Listening to the voices of *Wixage Anai* allows me to bring together the sensorial as well as political dimensions of indigeneity embodied in a radiophonic space, which is relevant for *Wixage Anai*, a radio program permeated by the voices, speech, and sounds of the late twentieth- and early twenty-first-century Mapuche movement.[5] My analysis emphasizes the multiple power relations that radio and sound symbolized and materialized in a colonial context, and, more broadly, in settings in which hegemonic and counter-hegemonic forces clash. The vocal impersonation of a character such as Indio Pije dispossessed Mapuche voices and racialized figurations—or rather disfigurations—of indigeneity via the settler Chilean mediascape, and this trend continued into the late twentieth-century Chile. Large media corporations took over the media and made it much more difficult for the Mapuche to gain a voice and audibility in society. Nevertheless, from the fringes of public space, the *Wixage Anai* team managed to foreground the culturally performative and politically effective force of a Mapuche radiophonic initiative. During its inaugural years, the program had significant reach, connecting with audiences across different regions of Wallmapu.[6]

Considering that Mapuche language, voices, and sounds had long been absent from the radio in Chile, to what extent did the *Wixage Anai* team's use of radio constitute an effective Mapuche linguistic, musical, cultural, and political interference in the hegemonic airwaves of acoustic colonialism? In this chapter I demonstrate that, during the 1990s and early 2000s, the radiophonic activism of *Wixage Anai* contributed to building public audibility and agency for the Mapuche in the domain of voice and sound. Their experience shows how radio and the mediascape can become spaces for Indigenous and oppressed voices to creatively generate Indigenous interferences and thus challenge regimes of acoustic colonialism.

Throughout the twentieth century, self-identified Mapuche voices were absent from the airwaves or, as we have seen, were merely "performed" as objects of mockery. *Wixage Anai* altered that hegemonic trend. Their irruption on the radiophonic ether—speaking the native language, playing Mapuche music, and embracing the broader territorial politics of Wallmapu—enabled the Mapuche radio team to exercise autonomy and agency within an "acoustic territory" (LaBelle 2010).[7] The Wixage Anai team was invested in formulating and practicing a Mapuche mode of radio that actively engaged the principles of nütram (conversation) and allkütun (attentive listening), especially during the 1990s and early 2000s. Paillan Coñoepan, who joined *Wixage Anai* around early 1995, describes this engagement in the following terms,

> El nütram es un elemento fundamental en la comunicación Mapuche, el conversar naturalmente en torno a un mate, o en torno a una comida o simplemente contarse historias, historias familiares. . . . El nütram es un elemento que caracterizó el programa *Wixage Anai* como parte de su fuerte. (Paillan Coñoepan, interview by author, 2023)

> Nütram is a fundamental component of Mapuche communications, the act of chatting naturally, around a mate tea, or around a meal or simply telling stories, everyday stories. . . . Nütram is an aspect that characterized *Wixage Anai* as part of its strength.

Paillan Coñoepan highlights that the program developed a way of broadcasting that was friendly to the Mapuche ear and captured the "attentive listening" of the audience, mainly members of the Mapuche community in greater Santiago. Moreover, along with cultivating nütram and inciting the exercise of allkütun, the radio show recovered a diverse range of Mapuche speech practices, such as *pentokun* (a traditional mode of greeting and self-introduction), *ngülam* (life advising), or open conversation programs directly produced from Mapuche homes, especially recording and broadcasting the voices of elders.

This Indigenous fight for voice and audibility drew me, as a Mapuche scholar in a North American university, to grapple with the positional and methodological challenges involved in interacting with the *Wixage Anai* team. From my very first encounter with the program's leadership team, I realized it was not enough to listen to the program from

FIGURE 4.1 · Promotional flyer for the radio show *Wixage Anai* at Radio Yungay advertising broadcast details and picturing members of the program team: José Paillal Huechuqueo, Elías Paillan Coñoepan, Ramón Francisco Curivil Paillavil, Ricardo Tapia Huenulef, and María Catrileo. Courtesy of José Paillal Huechuqueo.

the distance of the North and the insularity of the academy, essentially taking an intellectual stance from afar. Thus, in early July 1998, I accepted Curivil Paillavil's invitation to appear on the program about the emergence of a Mapuche literary scene. By that year, *Wixage Anai* had moved from Radio Nacional to Radio Yungay, a small station with a long history in Chilean radio, housed in modest studios in an apartment complex on Irarrázaval Street, a commercial area in the middle-class neighborhood of Ñuñoa in Santiago.

Entering the Radio Yungay studios and sitting in front of the main microphone was a momentous experience. Listening to Mapudungun in a radio booth, as the *Wixage Anai* hosts Elías Paillan Coñoepan and María Catrileo greeted me in our language, gave me a sense of reconnection to Ngulu Mapu and more broadly to Wallmapu. It awakened an intimate feeling of being part of the Mapuche of the South (Williche), native to the village of Tralcao, a colonized Williche territory in the southern Commune of San José de la Mariquina in Ngulu Mapu. This auditory experience was as memorable and moving as it was when

I first heard Mapudungun on the radio during my childhood in Tralcao. The experience of participating in a live radio broadcast in the Mapuche language, spoken by Mapuche voices, reframed how I situated myself as an engaged scholar in the struggle to make Indigenous voices audible.

Theorizations of "decolonizing methodologies" (Smith [1999] 2008) and "co-labor" (Leyva Solano and Speed 2008) help me establish a way to write about and engage with the radiophonic activism of *Wixage Anai*. Scholars Xochitl Leyva Solano (Mexico) and Shannon Speed (Chickasaw) state that "decolonization-oriented research in co-labor is not purely academic investigation, but rather it is born and reproduced in the interstices generated by the crossings of academia's others, flexible and open activisms, and social movements" (2008, 96). Based on the Mapuche principles of nütram and allkütun, and inspired by engaged research politics, over the years I have collaborated with the *Wixage Anai* team through politico-cultural dialogue and material support. Between the late 1990s and late 2010s, I became active in several initiatives, ranging from organizing a hemispheric gathering on Native uses of radio in Santiago in October 2006, to small fundraising campaigns and the purchase of technological tools, such as microphones and voice recorders. This personal involvement redefined how I related to the politics of decolonizing methods in grassroots Mapuche and anticolonial and emancipatory practices. My contact with the *Wixage Anai* team in 1998 generated a lasting and profound sense of Mapuche reconnection that has enriched what it means to be an engaged, attentive listener across the diasporic North/South distance. The principles of nütram and allkütun have informed how I relate to the *Wixage Anai* team members throughout my discussion and analysis of their radio experience. Cultivating conversation and attentive listening in the process of learning about the history and practice of *Wixage Anai* has been critical for grasping what a Mapuche politics and aesthetics of communications entails.

WAKE UP! GET UP!

Between 1993 and the early years of the twenty-first century, the transmission of *Wixage Anai* took place on several different radio stations in greater Santiago. The program moved from place to place as the broadcast team dealt with the ongoing challenges of finding an accommodating home and renting airtime with their limited financial resources. Its first radio home, from the program's June 1993 founding to 1996, was Radio Nacional Chile (CB 114 AM), where it ran on Saturday nights with the leading voices of Clara Antinao Varas and Curivil Paillavil at the microphone. The program was funded by the Divine Verb Congregation, a Catholic missionary congregation in the Florida Commune, eastern Santiago, which had offered their support thanks to a relationship Curivil Paillavil had established with them. When they received the congregation's funding offer, the *Wixage Anai* team insisted on maintaining control over the radio show, which would be fully Mapuche in content and format, without intervention from the sponsoring institution. The Divine Verb priest agreed to support and respect that principle.

Radio Nacional de Chile was a state-sponsored AM station with the broadest coverage in the country.[8] The Mapuche radio collective reached a wide audience across different regions in Ngulu Mapu, and sometimes even on the other side of the Andes Cordillera, into Puel Mapu. Audiences throughout the whole historic Mapuche territory of Wallmapu could listen—a considerable advantage of broadcasting from Radio Nacional. Unfortunately, relations with Radio Nacional soured. Officials of the radio station behaved rudely on the air toward Augusto Aillapan, a Mapuche spiritual and medicinal authority (Machi) and prominent member of the community in Santiago, bringing to a head the racism that the Mapuche broadcasters had frequently encountered at the station.[9] In July 1996, they decided to move to the smaller Radio Yungay.

One of the oldest radio stations in Santiago, Radio Yungay (CB 146 AM) had a more limited range that did not extend past the metropolitan area of the city. But Radio Yungay offered a less racist environment and charged less for airtime. Now, Wixage Anai could stretch Divine Verb's financial support to increase its frequency to one hour a day, Monday to Saturday, with Elías Paillan Coñoepan and María Catrileo at the helm. In 1999, *Wixage Anai* was nominated for most innovative radio show in Chile's Association of Entertainment Journalists (APES) annual award competition.

The program's next move, in September 2000, was to Radio Tierra (CB 130 AM), a feminist community-oriented radio station, where it remained until September 2013. Radio Tierra offered the *Wixage Anai* team airtime at a rate comparable to Radio Yungay's. But in addition, this station—with its investment in promoting women's political and cultural agency, human rights, and minority voices in Chile—had the great advantage of connecting *Wixage Anai* to a politically engaged community of non-Indigenous listeners in the greater Santiago area.[10] *Wixage Anai* spent thirteen fruitful years at Radio Tierra before switching to another old station in Santiago, Radio Panamericana (CB 142 AM). Other members of the collective relocated to the Mapuche-Lafkenche area in the province of Cautín, southern Chile.[11] With the financial support of just a few donors and sporadic community fundraising, airtime had to be constricted to one day a week. As a Sunday show, mainly under the leadership of José Paillal Huechuqueo, Margarita Elizabeth Huenchual Millaqueo, and Javier Salazar Cuminao, it continued on the air from December 2013 to December 2019. During these final years on the radio, *Wixage Anai*'s reach suffered from Radio Panamericana's extremely limited range and the scheduling disadvantage of airing at the same time as evangelical preaching shows. After its time at Radio Panamericana, *Wixage Anai* formally ended as a radio program.[12] In all the years that the program was on the air (1993–2019), *Wixage Anai* members worked as volunteers, devoting their time to planning, producing, and broadcasting the program without any compensation. They were driven by the aspiration to use Indigenous radio as a tool to contribute to and support Mapuche collective resurgence.

Chile was changing throughout the period *Wixage Anai* ran. When the program first went on the air in June 1993, hegemonic Chilean society was going through a "democratic transition" following Pinochet's seventeen-year dictatorship. This "transition" was characterized by a mixture of continuity and social fashioning of the neoliberal economic model of the Pinochet years. For the newly elected government of Patricio Aylwin (1990–94), this included redefining relations between the Chilean state and Indigenous peoples in a new "intercultural" policy directive articulated in the Indigenous Law of 1993, which aimed to add an "ethnic" ethos to the "Chilean nation" to make the neoliberal democratic regime more inclusive and representative.[13]

Parallel to the government's assimilationist interculturalism, an Indigenous political, cultural, and social movement was gaining strength, guided by the emergence of organizations and leaders who embraced the Mapuche's right to be and act as a nation autonomous from the Chilean nation-state. In 1989, the Mapuche organization Aukiñ Wallmapu Ngülam / Consejo de Todas las Tierras (Council of All Lands) was founded in Temuco and rapidly gained visibility and influence. In featuring the term *Wallmapu* in its name, the Consejo foregrounded a spatial concept that marked Mapuche land politics. In a geographical and historical sense, Wallmapu denominated All Lands of the Mapuche across Chilean-Argentine state borders. It was strategically translated as "Mapuche Nation." Echoing this land-based resurgence, organizations from the Mapuche urban diaspora, such as Meli Wixan Mapu, were also founded in Santiago in the early 1990s. The articulation of a visible and audible Mapuche movement coincided with Indigenous mobilizations across the continent, many provoked by the official "celebrations" of the five-hundred-year anniversary of the Spanish arrival in what came to be called the Americas. At this historical juncture, a hemispheric movement of Indigenous peoples raised their voices to express anticolonialism and embrace the banners of autonomy and self-determination.

The timing of its debut lent a particular resonance to the voices of *Wixage Anai* and its unexpected interference in airwaves that had been dominated by the metanarrative and mega-acoustics of the Chilean nation. As an independent Native media initiative, rooted in a sonic contribution to the reconstruction of Wallmapu, the Mapuche radio show disrupted the assimilationist script of the official Indigenist "intercultural" policies of the Chilean state by articulating an aesthetics and politics of the Mapuche as a people with their own language, knowledge system, history, and geography. Because of its success at generating independent funding, the Mapuche radio program maintained its commitment to promoting Mapuche views autonomously, guided by their relationships within the Mapuche movement. Taking advantage of Radio Nacional de Chile's transmission power, *Wixage Anai* committed to report on and amplify the voices of a Mapuche movement that, by the early 1990s, was turning to political mobilization in rural and urban communities, language revitalization initiatives, and multiple literary, artistic, and cultural expressions. The radiophonic space created by the *Wixage Anai* team celebrated Mapuche culture, revitalized Mapudungun, and supported the Mapuche movement.

Wixage Anai opened with a musical reference to the Mapuche people. The sounds of the trutruka wind instrument and the kultrun are linked to ceremonies and community events and to the political marches of contemporary Mapuche demonstrations. They unmistakably announced a Mapuche presence. This effect resonated across the airwaves to gather a Mapuche audience dispersed throughout greater Santiago and, thanks to recorded broadcasts relayed on local stations in southern Chile, in cities and rural areas of the historic Mapuche territories and beyond. The tapping of kultrun added a declaration of self-affirmation and strength. Then the kull kull and the kultrun signaled the start of *Wixage Anai* and the dissemination of Mapuche signs on the airwaves.

This introduction became a recognizable call to Mapuche and non-Mapuche listeners to gather around their radios. When *Wixage Anai* aired for the first time and the kull kull was heard, the Mapuche instrument that brings people together resounded and reverberated on many Mapuche ears. Elías Paillan Coñoepan, who joined *Wixage Anai* in 1996, described how this sonorous opening sent a clear message, "we were calling the people" (interview by author, 2023). A critical aim of the radio show was to attract a mostly urban Mapuche audience who were coming home after work, school, or from other venues flush with the noises of the dominant Chilean settler social environment. Hearing the kull kull and the kultrun on the air, all the senses, bodies, and spirits of a community tuned in. Their ears perked up. The Mapuche *trawün* (gathering) could begin, in an acoustic space made possible, ironically, through the technological infrastructure of the settler wingka society. Radio became indigenized. Despite the Mapuche dispersion and diasporization in urban settings, and the distance between Mapuches in the cities and those in rural areas of Ngulu Mapu and broader Wallmapu, radio transmuted into a connective medium.

From the outset, *Wixage Anai* enabled the Mapuche to feel a sense and sound of peoplehood in the radiophonic ether. Part of this poetics of convoking radio listeners and generating commonality inheres in the very name of the program, with its interjective force. Although the expression "Wixage Anai" is not generally written with exclamation marks, due to the absence of such graphic signs in Mapudungun grammar, the expression had more force on the air than it had on paper. At the beginning and end of the radio show, the Chilean broadcaster and Mapudungun native speaker Clara Antinao Varas vocalized with interjective and vocative power: *Wixage Anai!*

To the Mapuche ear, this Mapudungun utterance has an expressive vigor which, according to Antinao Varas, evokes the power of Mapuche oral discourse in the past. She points out that "in old times, the speech of traditional authorities such as the *pu longko* [representatives of communities-territories] was very energetic" (Antinao Varas, interview by author, 2012). In the absence of technologies of sound amplification, a loud voice together with a richness of tones and phonetic turns was key to animating audiences in large community gatherings. As linguist María Catrileo writes, "Interjections were, in past times, frequently used in a speech act in Mapudungun" (2010, 13). Echoing this discursive tradition, the expression Wixage Anai, voiced emphatically, was immediately understood in its multiple meanings as an identity marker of the radio show, as a communication initiative calling on the Mapuche "to get up"—to arise—in times of resurgence and collective struggle. The mere act of intoning "Wixage Anai" was already an interference in the vast space of the airwaves as it traveled to the intimate space of listeners' ears. With a touch of humor, Antinao Varas noted parenthetically that, given that the radio program was broadcast at night, saying its name—Wixage Anai—was also a humorous command to stay awake: "We told the Mapuche radio listeners, in Mapudungun, 'Do not fall asleep yet'" (interview by author, 2012).

LANGUAGE SOUNDS

In this radiophonic Mapuche awakening, employing Mapudungun was critical. Although announcers sometimes alternated between Mapudungun and Spanish, the voices of Curivil Paillavil (RCP) and Clara Antinao Varas (CAV) prioritized Mapudungun to make sure *Wixage Anai* altered the linguistic texture of Chilean radio. The announcers' Indigenous linguistic and sonic inflection was evident in the inaugural show as the voices of Curivil Paillavil and Antinao Varas went live on air:

> RCP: Mari mari pu peñi, pu lamgen, kom pu ce ajkvtulelu tvfa ci zugu. Mari mari, lamgen. (Hi brothers, hi sisters, and hi to all the people who listen to this program. Hi, sister Antinao.)
>
> CAV: Mari mari, lamgen. Mari mari pu lamgen, pu weni, kom pu ce, petu ajtukulu tvfa ci zugu. (Hi, brother Curivil. Hi to all sisters and brothers, friends, and all people who are listening to this program.)
>
> RCP: Hola, qué tal amigos. Con este saludo iniciamos nuestro primer programa Mapuche que lleva por nombre *Wixage Anai*, que significa: "Álzate y Despierta: Camina con Dignidad." Ante el micrófono, están con ustedes, Clara Antinao y, quien les habla, Ramón Curivil. (Hi, how is it going, friends? With this greeting, we

begin our first Mapuche radio show, which is named *Wixage Anai*, which means "Get up and wake up; Walk with Dignity." On the mike, here with you all, are Clara Antinao and me, Ramón Curivil, speaking to you now.)

CAV: Felei, pu lamgen. Faci antv mai, . . . jituayu tvfa ci zugu. Welu ta Mapuce gijatukei wvne, fei mew ta kvme amukei zugu ka kvme ta xipakei. Wvne pici gijatuaiñ mai pu lamgen. Eimvn ta keyumuaiñ akupe ta mvn heweh. (Okay, you all. This day, yes . . . this program. But first, to gather the Mapuche, may this go well, and may it get out well; first, it is a little gathering for all the Mapuche. That you all are taking care to achieve good strength for all of us.)

RCP: Feley, lamgen . . . (Okay, sister!)
(*low voices follow*)

CAV: Wvh fvca, wvh kuse keyumuaiñ tvfa mew . . . (Ancestral man of the dawn, ancestral woman of the dawn, help us in this program . . .)

RCP: Wenu fvca, wenu kuce faciantv mai xipaiai ta iñ zugu, elumuiñ kvmeke rakizuam, kvmeke zugu; fei mai caw kvme xipaiai ta iñ zugu. (Ancestral man from above, ancestral woman from above, may our program go well today, give us good thought, also good matters; ancestral father, may our program go well.)

CAV: Jituaiñ kiñe we kvzaw tvfa, kvmeke am ta akupe, kvmeke rakizuam ta akupe, fei mew ta kvme xipaiai ta iñ zugu, felepe mai pu lamgen, pu papai. . . . (We start a new work now, for a good outcome, for good thoughts, for a good program, ready for you all sisters and brothers, and elders. . . .)

RCP: . . . (*murmurs in Mapudungun, then returns to a regular tone*) . . . Felepe mai! Felei pu peñi, felei pu lamgen, zew pici gijatuiñ, fewla mai, amuleai ta iñ zugu! (. . . So be it! Okay, brothers, okay sisters, our brief prayer is done; now, yes, let's move onto our subject!)

Curivil Paillavil and Antinao Varas's linguistic interventions typically began with the standard Mapudungun greetings: "Mari mari . . ." Then Curivil Paillavil transitioned into Spanish to greet the audience and translate the meaning of Wixage Anai: "Get up and wake up: Walk with Dignity." Immediately afterward, he introduced the main voices at the microphone. Next, in Mapudungun, Antinao Varas wished the best for the radio program. Finally, switching to a softer voice, her speech took the form of a *llellipun*—a traditional prayer for good omens characteristic of Mapuche protocols to inaugurate a meeting, work, or other event. By the end of the segment, Curivil Paillavil's voice would return to normal volume as he addressed the audience in a conversational tone: "Felei pu peñi, felei pu lamgen, zew pici gijatuiñ, fewla mai amuleai ta iñ zugu."

One of *Wixage Anai*'s traits was their familiar and colloquial tone of voice, which was consistent with the politics of using radio according to the Mapuche principle of nütram. As Paillan Coñoepan told me, "The idea was not to put on FM-stereo-style voices or voices of 'professional' broadcasters; rather, the objective was simply to reach people as one speaks, a colloquial language, but also in Mapudungun, as our people speak, as they express themselves daily. . . . We the Mapuche, especially in Mapudungun, have a tonality, a linguistic characteristic that is part of our own culture, of our language, and our idea was to recuperate that too and to bring it onto radio" (interview by author, 2023). Coincidentally, in the late twentieth-century mediascape of Chile, a new generation of radio broadcasters and radio shows was also shifting to more colloquial ways of speaking before the microphone, abandoning the previously prevalent style of the *voz impostada*, that is, speaking with an affected, "imposter" voice. The nütram aesthetics and politics Wixage Anai put in practice coupled well with the stylistic turn that was reshaping the contemporary radio scene.

Hearing the native language marked a pivotal moment for Mapuche radio listeners in the greater Santiago area as well as in more rural southern Chile. Evoking the impact of this linguistic inflection, Margarita Elizabeth Huenchual Millaqueo, who joined the program in the late 1990s, concluded, "Many people spoke the Mapuche language again thanks to radio" (interview by author, 2018). Here she was referring to the elders of the large Mapuche community, native speakers of Mapudungun, who, since their migration from Mapuche rural communities to peripheral areas of the metropolitan region, such as Cerro Navia, La Florida, Peñalolén, El Bosque, La Florida, and San Bernardo had repressed their knowledge of the language and avoided using it. Recounting her own similar relation to language, Huenchual Millaqueo told me, "A while ago, I felt embarrassed speaking Mapudungun" (interview by author, 2018). She attributed hearing Mapudungun on *Wixage Anai* to counteracting these feelings and contributing to a valuation of the Mapuche language. Employing Mapudungun equally with Spanish in radio communication legitimized its use on the radio and generated public audibility for an Indigenous language in Chile.

Huenchual Millaqueo's experience was far from singular. Putting Mapudungun on the air interfered with the script of linguistic and phonetic colonization in Spanish, which has been historically established through the school system, mass media, and the disciplining of public spaces. The history of this colonizing process was long and painful for Mapudungun speakers, especially in urban settings where they had been silenced for decades. In the prologue to her *Diccionario ta iñ mapun dungun: Nuestra lengua Mapuche* (2014), Antinao Varas provides a similar account of conflicting feelings about language following her migration to the capital in the early 1960s. She recalls how attending the gatherings in Santiago's Quinta Normal Park, a fixture of Mapuche life since the mid-twentieth century, contributed to a vexed sense of Mapuche identity. There, she experienced the practice of fellow Mapuche silencing their own native language in this public space.

> Allí siempre iba con la intención de hablar el Mapudungun con alguien, pero, me equivoqué. Nadie quería hablar en su idioma Mapuche. Con la negativa de mis hermanos y mis hermanas para hablar en su lengua, no fui más. (Antinao Varas 2014, 7)

> Üye meu rumel küpa Mapudungunkefu inei iñchuu-rume, welu welulkawün: ineino-rume küpa dungulai ñi Mapunche Dungun meu. Pinulu ñi pu lamngen dungualu ñi dungun meu, amuwelan. (Antinao Varas 2014, 12)

> I used to go there with the intention of speaking Mapudungun with someone, but I was mistaken. Nobody wanted to speak in their Mapuche language. With my Mapuche kin refusing to speak the language, I stopped going.

Antinao Varas's story testifies to self-silencing as a Mapuche mode of survival amid the acoustic colonialism they experienced in an urban setting. Migrants to the capital were quick to sense that it was only socially acceptable to speak the colonial language, Spanish, or European languages with cultural "prestige." To speak Mapudungun was to risk racial, linguistic, and cultural discrimination. Given the coupling of linguistic subjugation and racial discrimination, the use of Mapudungun inevitably declined, and Spanish became the language of choice. Recent studies on the state of the native language in the early twenty-first century show that members of the younger generations do not widely speak it. Mapuche linguist María Catrileo asserts,

> Aun cuando la lengua ha sido capaz de sobrevivir durante el transcurso del tiempo, no se puede sostener que este medio de comunicación constituye una lengua saludable. Según las investigaciones realizadas por organismos públicos y privados, existe un número cada vez más reducido de hablantes. La mayoría de ellos son adultos mayores o niños que están siendo incentivados por sus propios abuelos para aprender la lengua de los Kuyfikeche o ancestros. (Catrileo 2010, 40)
>
> Although the language has managed to survive through time, it cannot be said that it is a healthy language. According to research conducted by public and private organizations, there are fewer and fewer speakers. Most are older adults and children whose grandparents encouraged them to learn the language of the *Kuyfikeche* or ancestors.

Submission to the colonial script clearly became an obstacle to the efforts to reestablish Mapudungun in public spaces and even, in many cases, in Mapuche homes. In this context, speaking and hearing the native language on the radio was a powerful act of Mapuche resonance in the media, underscoring the language's linguistic, communicational, and aesthetic value and asserting its political relevance for the contemporary Mapuche movement. Furthermore, for written communications and documentations, such as those published by the Jvfken Mapu Center, the radio team chose to use the Mapudungun grammar or alphabet created by the Mapuche independent scholar Anselmo Raguileo in the mid-1980s.[14] Most Mapuche grassroots organizations had adopted this graphemics system, which reflected a sense of Mapuche autonomy by aiming to closely follow the phonetics of Mapudungun in writing. Through this linguistic practice, the *Wixage Anai* collective complemented their radiophonic stance that the Mapuche have their own language and constitute a people.

While the speaking and dissemination of Mapudungun through radio demonstrates radio's potential to promote the value and enhance the sonority of an Indigenous language, it is also unexpected and paradoxical. As a language symbolically associated with nature, "the language of the earth" resurges through the technological assemblages between the control room and the sound booth. Mapuche sounds and voices go "on the air" amid microphones, electronic levers, artificial lights, and acoustic devices.

Interactions with listeners happened only through the telephone and letters. But, by the time the program came to the feminist station Radio Tierra (2000–2009), electronic and digital devices facilitated communications with listeners through messaging apps and email. The urban and radiophonic experience of *Wixage Anai* set in motion creative and hybrid crossings between the medium of orality in modern Western society (radio transmission) and the orality of ancestral speech traditions. This interweaving reflects the creative ability of the Mapuche who seek to leave their acoustic and semantic echoes in the ears of a contemporary social body.

The *Wixage Anai* team had aimed from the start to engage non-Mapudungun-speaking audiences. They included Spanish translations in the script in hopes of reaching those who had lost their connection to the language, particularly in the context of a diasporic and urban audience in greater Santiago. Spanish translations, however, also helped *Wixage Anai* attract and build bridges with non-Mapuche radio listeners and forge alliances over the acoustic territory. From the first broadcast, Antinao Varas and Curivil Paillavil alternated between Mapudungun and Spanish:

> RCP: Estamos iniciando nuestro primer programa. El propósito que nos anima es acompañarle todos los fines de semana por un espacio de media hora, con el fin de animar y promover el desarrollo de nuestra cultura en sus diferentes aspectos. Este será un programa que haremos en bilingüe, debido a que nuestros hermanos aquí en Santiago la mayoría no habla el Mapudungun. *Felelai, lamngen?*
>
> RCP: We are launching our first program. Our animating purpose is to accompany you for a half an hour every weekend to encourage and promote the development of our culture in its different aspects. This will be a bilingual program, for our brothers and sisters here in Santiago, most of whom do not speak Mapudungun. *Agree, sister?*

CAV: Felei pu lamgen. Inciw ta mvlepaiayu kom Sábado antv fante pun, epu rume zugun mew zuguayu Mapuzugun mew ka wigka zugun mew. Faw ta iñ pu weceke ce ka ta iñ pu weni kimlai Mapuzugun egvn, fei mew ta epu zugun mew ta zuguayu. Kom zugun kvmei nvxamkan mew, pu lamgen. Ta iñ kuyfike ceyem, re Mapuzugun mew elzugukefuygvn.

CAV: Okay. We will be here every Saturday night, both of us speaking in Mapudungun as well as in the wingka language. Now for all the young people and allies who do not know Mapudungun we will translate it for you. All languages need to have a good conversation. For our elders, Mapudungun will be spoken as well.

In the Spanish segment, voiced by Curivil Paillavil, the reiteration of the idea of *animar* (to liven up or encourage) emphasizes the goal of stimulating the Mapuche—*nuestros hermanos* (our brothers and sisters)—who do not speak the language to connect. Thanks to this translation to the wingka language, those radio listeners become equally immersed in Mapudungun. At the end of his statements in Spanish, Curivil Paillavil uses an expression in Mapudungun, "Felelai, lamngen?" (Agreed, sister?) to signal that translation is a linguistic and gender transition. They reiterate this speech pattern as their script evolves:

RCP: Este programa va dirigido especialmente a ti peñi, a ti lamngen, que conoces la dureza de la vida santiaguina y que por la influencia de los medios de comunicación, especialmente la radio y la televisión, te hace olvidar muy luego de tus valores Mapuche. Pero también queremos dirigirnos a nuestros hermanos del campo que luchan, como nosotros, para mantener su dignidad, su identidad, a pesar de todas las dificultades. *Feley ka, felelai, lamngen.*

RCP: This program is especially addressed to you, peñi [brother], for you, lamngen [sister]. You know how difficult life in Santiago is. The media, especially radio and television, have made you quickly forget your Mapuche values. But we also want to address our brothers and sisters in the countryside who, like us, struggle to maintain their dignity, their identity, despite all the difficulties. *All right, all be right, sister.*

CAV: Feleyei mai lamgen. Pu lamgen, pu papai, pu caw. Eimvn mew kvpa puwiiñ. Eimvn kvme kimnieimvn cumgeci ta iñ kvxankawken tvfa ci fvta waria mew. Faw ta mvletuy re kake ce ta ñi zugu, re wigka ta ñi az fei mew ta kiñeke mew gujikeiñ ta iñ Mapucegen. Eimvn kai pu lamgen lelfvn mew mvlelu, ka kiñeke mew ce xokigekelaimvn ka femgeci kuxankawkeimvn inciñ reke.

CAV: Okay, brothers, female elders, parents. You all want to be happy. You all will be having conversations in Mapuche here in the city. Now . . . the non-Mapuche influences make you forget your Mapuche being. To all the Mapuche kin in the countryside, here we are, and as one people we are here, too.

Curivil Paillavil and Antinao Varas make explicit that their primary audience is Mapuche residents of Santiago. Several studies reveal that a high percentage of the Mapuche people live in the cities of southern and central Chile, particularly in Santiago, due to what Mapuche educator Felipe Curivil Bravo calls the "Mapuche exodus" (2012, 165). According to him, this displacement of Native families and people had "its origin in determining structural factors," such as "the demographic pressure upon Mapuche *reducciones*, small land spaces to which the Mapuche people were confined" during and after the late nineteenth- and early twentieth-century Chilean colonial invasion, occupation, and settlement of ancestral territories (Curivil Bravo 2012, 186). Mapuche researcher Enrique Antileo Baeza similarly points to Chilean colonial history as a constitutive aspect of "the structural factors of domination" that triggered the arrival of the Mapuche population in cities within Wallmapu or outside it (such as the case of Santiago)" (2012, 193). This forced migratory movement paved the way for the formation of "the concept of diaspora" (Antileo Baeza 2012, 197–200). Challenging the urban/rural dichotomy, Curivil Bravo and Antileo Baeza understand the Mapuche community-formation process in cities within a broader history of Native removal from ancestral territories.

The *Wixage Anai* team focused on the displacement and hardship that Mapuche families and people underwent, the "difficult life in Santiago," while settling in the capital. Curivil Bravo, in his compilation and study of testimonials from Mapuche immigrants in Santiago in the early 1960s, describes the "labor precarity of the Mapuche" who had to take "less skilled jobs," which exposed them to "a tense work environment in which, to survive in the city, the Mapuche must endure disturbing racism." These conditions illustrate "the economic dimension of the situation of colonial subordination, in which class exploitation and everyday racism are sides of the same coin" for Mapuche immigrants in the wingka city (Curivil Bravo 2012, 172). The "concealing of cultural features" becomes "one of the first responses of the Mapuche on arrival in Santiago," which results in a traumatic "psycho-social internalization and reproduction of colonialism" (Curivil Bravo 2012, 166).

Curivil Bravo's study explains and confirms the embarrassment or pervasive denial of origins that Antinao Varas, like Curivil Bravo himself, described as "concealing." Her eagerness to connect with fellow Mapuche on Sundays in the Parque Normal of Santiago was squelched when she saw that "the Mapuche gathered there, many Mapuche; you could see five hundred Mapuche walking, chatting . . . but in Spanish." Although the Parque Normal was a social gathering place, people avoided coming out as Mapuche: "They felt ashamed of being Mapuche. . . . I did not meet anyone willing to speak the language." Still worse, Antinao Varas added, "I tried to speak in Mapudungun with them. . . . They laughed at it" (interview by author, 2015).

Curivil Bravo notes that, for the Mapuche who migrated to cities like Santiago, cultural "concealment" was a way to endure an aggressively assimilationist environment (2012, 187). Thus, many Mapuche, in their struggle to survive the Chilean settler city, tended to avoid speaking Mapudungun in public. Antinao Varas's narrative is a clear account of how removal from Ngulu Mapu led many Mapuche to hide in the wingka language to survive the acoustic colonialism that framed their linguistic and social life in the Chilean city. Awkward laughter emerged as a mode of simulation, a gestural and corporeal disguise. It is a mark of the trauma from feeling unable to speak the native language and finding oneself estranged when hearing it. Whether imposed from within or without, such censorship impacts Mapuche racial, economic, social, and cultural survival in the Chilean capital. The launching of *Wixage Anai* in June 1993 directly impinged on this colonial pattern of verbal and aural subordination and challenged the logic of concealment or self-concealment. As a staging of Indigenous agency in the enunciative realm, even from the margins, *Wixage Anai* altered the hegemonic media ecology of the period.

Mega-corporations overwhelmingly controlled mass media by the 1990s, disseminating Chilean society's hegemonic settler capitalist, colonial script. This corporate turn was detrimental not only to Mapuche voices in the mediascape but also to local communities and grassroots movements. Mass-media outlets were largely inaccessible to the autonomous voices of the resurging Mapuche movement that began speaking in the public domain in Chile in the 1990s. In *Los magnates de la prensa: Concentración de los medios de comunicación en Chile* (2008), a detailed study of the corporate concentration of media in Chile, researcher and journalist María Olivia Mönckeberg describes this hegemonic reordering:

> En los noventa llegaron empresas extranjeras con filiales en otros países latinoamericanos a la conquista de radioescuchas chilenos. Al mismo tiempo, los grandes grupos locales dueños de diarios decidieron ampliar su marco de acción, y se fortalecieron algunas cadenas que asfixian las numerosas iniciativas regionales que caracterizaban hasta hace poco el variopinto panorama radial chileno. En 1999, estos grandes consorcios representaban el 31 por ciento de las emisoras del país, con un total de 308 radios. (Mönckeberg 2008, 370)

> With branches already in other Latin American countries, in the 1990s foreign companies arrived to conquer Chilean radio listeners. At the same time, the large local newspaper owners' groups decided to expand their radius; and some media chains strengthened such that they strangled the regional initiatives that characterized the multicolored Chilean radio scene. In 1999, these large consortia represented 31 percent of the radio stations in the country, within a total of 308 stations.

The neoliberal corporatization and transnationalization of the media altered radio.[15] Previously, the Chilean radio scene had been characterized by its "multicolored" panorama, with a wide variety of local stations across provinces and regions. Certainly, from a "Chilean" standpoint, this offered a diverse range of sonorities. Nevertheless, from a Mapuche perspective, Indigenous Peoples practically did not exist in the linguistic, musical, cultural, and news script of that "multicolored" radiophonic scene. In the 1990s, there was no precedent for a radio station owned, controlled, or managed by Mapuche individuals or collectives. The mega-corporate establishment structure aggravated the situation. Possibilities for autonomous access to Indigenous radio became distant. Meanwhile, corporate, neoliberal Chile's deployment of colonizing sonic waves of strengthened over Ngulu Mapu's "acoustic territory" (LaBelle 2010).

Even as a tenuous yet dissonant interference, *Wixage Anai* marked an inflection within this new mediascape. The Mapuche radio team emerged from the margins of the hegemonic urban setting when it bought airtime at Radio Nacional de Chile under terms that allowed it to exercise an autonomous politics of communication. Radio Nacional was in its final years as a state-owned station and the democratic transition fostered a climate of relative acceptance of a Mapuche program. *Wixage Anai*'s urban radiophonic experience set in motion hybrid crossings between an influential technology of orality in Western society (radio transmission) and the orality of ancestral speech traditions, an interweaving of codes and techniques of communication that reflects the creative ability of the Mapuche to navigate diverse mediascapes.

In retrospect, we see that the Mapuche use of radio constituted an effective mode of Indigenous interference and anticolonial activism in the territory of broadcasting and listening. Radio became an instrument to amplify the language and reconstruct the public audibility of the Mapuche. It succeeded in its effort to interfere in the radiophonic ether of the wingka city and the global airwaves.

ÜLKANTUN: MAPUCHE CHANT RESURGES

In a 2012 interview, José Paillal Huechuqueo, one of the younger members of the foundational team of *Wixage Anai*, remarked that he was always concerned with the role of music in the show. Paillal Huechuqueo told me about the difficulties the team confronted when trying to include music played on Mapuche instruments and lyrics in Mapudungun. Armando Marileo, a *kimche*—a Mapuche devoted to cultivating and protecting the memory of Native knowledges—with whom the team collaborated, had produced Mapuche music recordings by the early 1990s for his own radio initiative in Temuco, a city at the center of Ngulu Mapu. Recalling their deficiencies as well as Marileo's support during the first months of *Wixage Anai* on Radio Nacional, Paillal Huechuqueo said,

> Lo que recuerdo harto de ese tiempo es que no había música Mapuche, . . . no había música Mapuche. Entonces recuerdo que en uno

> de los primeros programas nosotros pusimos música de Illapu. Era lo más cercano al tema Mapuche, un tema que tenían ellos y uno que otro ülkantun de un cassette. . . . No sé de dónde lo sacamos, pero recuerdo también que sacamos música del mismo programa de Armando Marileo. . . . Entonces una de las preocupaciones de nosotros era, si pensábamos seguir haciendo el programa, teníamos que generar música Mapuche propia. (Interview by author, 2012)

> What I remember well from that time was that there was no Mapuche music, . . . there was no Mapuche music. Then, I remember that on one of our earliest programs, we played music from Illapu. It was the closest we could get to a Mapuche song, a song that they had as well as a few ülkantun on a cassette. . . . I don't know where we got it, but I also remember that we got music from Armando Marileo's program. . . . And then one of our concerns was, if we planned to continue making the program, we had to generate our own Mapuche music.

There were almost no recordings of Mapuche chants and music—another sign of the wingka domination of Chilean culture in the 1990s. Lacking recorded Indigenous musical sources led the *Wixage Anai* team initially to play songs in Spanish by Chilean bands such as Illapu, a group linked to the New Chilean Song, a politically engaged musical current of the 1970s. Besides their commitment to social movements, most of Illapu's songs are characterized by the incorporation of wind instruments from the Quechua and Aymara traditions, such as *quena* and *zampoña*, mixed with Western instruments. Although it was not Mapuche, the *Wixage Anai* broadcasters resorted to this music because it at least fit into the program's political objectives and sonorous script. Furthermore, Illapu had consistently expressed solidarity with the Mapuche movement. But the limited availability of musical sources from Mapuche composers and chanters, with lyrics in Mapudungun and based in Mapuche aesthetics, compromised the show's objectives. Given the *Wixage Anai* collective's awareness of the importance of practicing an Indigenous politics of communication that was Mapuche in content and in form, they searched for Indigenous music for the show, but it was a major challenge, as Paillal Huechuqueo explained to me:

> Hacer comunicación Mapuche no tiene que ver con sentarse y "hablar de" lo Mapuche, porque eso puede hacerlo cualquiera que estudie.

> Hacer comunicación Mapuche es elemental hacerlo en Mapuche y desde lo Mapuche. Y, lo mismo, hacer música Mapuche no tiene que ver con tocar instrumentos Mapuche. Tiene que ver con hacerlo desde lo Mapuche, . . . desde las formas, desde la vivencia, desde el sentido que se le da al sonido, a las palabras, en el mundo Mapuche. (Interview by author, 2012)

> Making Mapuche communication involves more than sitting down and "talking about" Mapuche things, because anyone who studies can do that. To make Mapuche communication, it's fundamental to do it in Mapuche and from the Mapuche. Similarly, making Mapuche music isn't about playing Mapuche instruments. It's about basing it on Mapuche ways, lived experiences, from the meaning given to sound, to the words, in the Mapuche world.

Paillal Huechuqueo suggests that radio communications, like music, are an art form. To produce and practice them, both the "what" and the "how" are important. Making radio and music with a "Mapuche communication" practice entails the challenge of doing it in "Mapuche ways" and "from the Mapuche" perspective. Thus, the interweaving of "forms" (aesthetics), "experience" (personal and collective history), and "meaning" (semantics) was essential for the poetics and politics of *Wixage Anai* and its goal of birthing a Mapuche radio show. This ethos positions aesthetics at the core of the conversation on indigeneity and the media. In the case of radio, it foregrounds the need to engage with Indigenous modes of sonority, including music and the native language. Thus, despite the limited existence of recorded music in Mapudungun in the program's early years, the *Wixage Anai* team set about collecting sources that matched their aesthetics and politics.

Indeed, at the inaugural program on June 26, 1993, the musical and vocal sounds of ülkantun—a Mapuche chant in Mapundungun—went on the air. As a chant, ülkantun is one of many variations of *ül*, the more generic term for Mapuche vocal, musical, and poetic arts. At *Wixage Anai*'s inaugural emission, the first ülkantun aired had been collected by Ramón Curivil Paillavil. Interpreted by a male voice, the chant was accompanied by the sound of a *pifilka* flute. It was a chant of love addressing the persona of Selinda, a female figure that has been a motif in many ülkantun and who symbolically embodies the beauty of Mapuche women. A second ülkantun by *ülkantuchefe* (Mapuche chanter) María Canio, from Metrenko, a community south of Temuco also aired. Delivered without instrumental accompaniment, Canio's chant tells a community story. Clara Antinao Varas then explained, "We the Mapuche tell history through singing, conversation, and story." She thus highlighted the significance of including ülkantun as an integral part of the sonic and cultural Mapuche politics of *Wixage Anai*.

Another ülkantun the *Wixage Anai* team aired in the late 1990s features a major figure in the Mapuche struggle against Spanish colonialism, Kajfvlikan, or Caupolicán, as commonly written in Spanish. Probably, the lyrics of "Kajfvlikan" was not understood by many Mapuche and non-Mapuche listeners; but, given the slow flow of its rhythm and its cadence in Mapudungun, many of them enjoyed it. Here I leave the transcription of the lyrics of "Kajfvlikan" untranslated, to invite you as readers to overcome the graphic mediation of writing and to *aurally imagine* how the Mapuche chant may have sounded on the airwaves,

> Wvle ci antv, wvle ci antv, wvle ci antv
> Kuzeialu, kuzeialu Kajfvlikan
> Wvle ci antv ga, wvle ci antv ga
> Kuzeialu ga Kajfvlikan
> Bubulmekei ñi, BuBulmekei ñi
> Yelkonal ga Kajfvlikan
> Bubulmekei ñi, BuBulmekei ñi
> Yelkonan ga Kajfvlikan
> Wvle ci antv, wvle ci antv
> Kuzeialu ga Kajfvlikan
> Wvle ci antv ga, wvle ci antv ga
> Kuzeialu ga Kajfvlikan
> Wvle ci antv naqpvrapaci antv mew ga
> wewmapuialu Kajfvlikan
> Wvle ci antv ga, wvle ci antv ga
> naqpvrapaci antv mew ga
> wewmapuialu Kajfvlikan
> Wvle ci antv ga, wvle ci antv ga
> majmatuiatuay kuzefe kai
> Wvle ci antv majmatuiatuay kuzefe kai
> wewgen em ga piawtuai ga kuzefe kai.

Even in the late 1990s, "Kajfvlikan" was one of the few recorded ülkantun that could be played on the radio. It came on a cassette titled *Mapuche*, issued in 1998 by the Santiago-based Alerce record label, which included a series of ülkantun. Delivered as an a cappella solo through the voice of a male ülkantuchefe, "Kajfvlikan" staged the vocality of the ül. As in María Canio's chant, the ülkantuchefe's voice stands out. In its paused voicing, the chant acquires a deep vocal and melodic resonance. Listeners could note the ülkantun's slow rhythm and cadence. Thematically, this ülkantun celebrates the virtues of Kajfvlikan. It is useful to recall here that his figure was archetypically aestheticized and iconized as a "warrior" in Ercilla y Zúñiga's epic poem *La Araucana*, written in the mid- and late sixteenth century. As I explained in chapter 1, this epic poem, canonized as part of "Chilean literature," instituted the term *Araucano* to represent the Natives of Ngulu Mapu in settler colonial discourse. Ercilla y Zúñiga's epic also profiles our Mapuche society as one dominated by male "warriors."

Against this archetypical representation, in the chant, the figure of Kajfvlikan becomes a more familiar and approachable persona, portrayed not as the iconic "Araucanian warrior" but as a leading player in one of the most traditional team games in Mapuche sports and social life, namely, *palin*. This game is a festive and recreational event in Mapuche social life where community members share food, music, and conversation. Called by his Mapudungun name, his persona is placed in a temporal line that borders the present, "Wvle ci antv . . ." (tomorrow). Kajfvlikan is no longer part of an archetypical idea of the past, or an abstract paradigm of the Western past-present-future logic. Resituated in the "Wvle ci antv . . . ," his persona is closer to the listener's times, like a Mapuche kin. The soothing repetition gives familiarity to this ülkantun, as these lyrics make clear:

Wvle ci antv, wvle ci antv, wvle ci antv
Kuzeialu, kuzeialu Kajfvlikan
Wvle ci antv ga, wvle ci antv ga
Kuzeialu ga Kajfvlikan

Tomorrow, tomorrow, tomorrow
he will compete, Kajfvlikan will compete
Tomorrow, tomorrow.
he will compete, Kajfvlikan will compete.

In the 1990s, the broadcast of this type of Mapuche chant into the radiophonic ether was an aural and sonic interference. First came awareness of a unique sound that broke from the Westernized musical soundscape; a break, an inflection, a sonic turn that interrupts the hegemonic waves of acoustic colonialism. The ülkantun has a rhythmical sonority lacking in the music played on Chilean radio. The inclusion of ülkantun infused *Wixage Anai* with what Paillal Huechuqueo called communication "in Mapuche" terms and "from the Mapuche" perspective. By airing Mapuche chants without translation, the program also encouraged its audience, especially speakers of the language, to practice allkütun—"attentive listening"—the cardinal concept of Mapuche listening and communication I have also been practicing throughout this book. The rhythmic voicing and musical pace of ülkantun urged the audience of *Wixage Anai* to listen acutely.

The slow cadence of the Mapuche chant helped the program cultivate allkütun among its Native and non-Native radio listeners and fostered their loyalty. In radiophonic culture, audience loyalty means "more than simply tuning in," they form "an attachment" that entails "attention and belief" on the part of those who listen (Bronfman 2016, 117). In the radiophonic practice of *Wixage Anai*, these dimensions of attention, loyalty, and affect became extremely relevant as the program invested in rebuilding a sense of Mapuche being in the urban setting of greater Santiago, regenerating affective and communicative ties between Mapuche across urban and rural areas, and reconstructing a sense of common belonging to Wallmapu. Attentive listening to the slow pace of an ülkantun was also part of a broader experience of Mapuche aural and temporal autonomy, while it unleashed a disruptive force over the colonized spaces of Chile/Wallmapu.

Facing the paucity of recordings of Mapuche music, throughout the 1990s and early 2000s, the founders of *Wixage Anai* expanded their repertoire of ülkantun by building their own musical archive. In the early years, Curivil Paillavil and José Paillal Huechuqueo visited the homes of Mapuche elders who cultivated Mapuche chant. They approached elders in Santiago and in Mapuche communities in Ngulu Mapu for help. It was no easy task. Curivil Paillavil recalls that in 1993, "It was very difficult to get interviews from people." However, as the radio show gained more and more listeners, "people lost the fear" of speaking (interview by author, 2022). Their person-to-person contacts, conversation, or what the Mapuche call nütram, were critical to building trust and support, and especially in persuading elders of the relevance of recovering musical traditions and Native knowledges to ensure the reconstruction and continuity of Mapuche lifeways. A highly respected member of the Mapuche community in Santiago, elder Augusto Aillapan, who was a Machi—a traditional medicinal and spiritual authority—agreed to record some chants for the radio program. Because their outreach to the elders put the radio team in touch with many previously unknown Mapuche musicians, in 1994 the *Wixage Anai* team, with the support of the Mapuche community grassroots organization Meli Rewe, organized the First Gathering of Mapuche Chant and Music. The event took place where most of the leaders of the Mapuche radio collective lived and where Meli Rewe was based, namely, the Commune of Pudahuel, in western Santiago.

The organizers held the Mapuche art and music festival in an elementary school in Pudahuel. During the two days of the gathering, the *Wixage Anai* team recorded interpretations of ülkantun by several elders, who happened to be mostly of Lafkenche background, from the coastal territories. Among the several ülkantuchefe who participated and were recorded in this community festival were Rosa Cuminao, Bartolo Malo, Alfredo Paillal, Leandro Painequeo, and José Painequeo. Following its success in 1994, the festival in Pudahuel was repeated annually for several years. For Margarita Elizabeth Huenchual Millaqueo, this gathering was pivotal in making Mapuche musical traditions audible for the community in Santiago, in addition to satisfying the objectives of the radio show. "It was like an awakening for the Mapuche; the Mapuche loved their language again, loved their own chant again, their own knowledges" (Huenchual Millaqueo, interview by author, 2018).

The sound of the native language and music enabled *Wixage Anai* to fulfill its aim of injecting the radio waves with a spirit of Mapuche self-affirmation. Speaking Mapudungun and playing Mapuche music were critical dimensions for *Wixage Anai*'s foundational goal, namely, "to animate Mapuche lifeways in the city and in the country" (Wixage Anai 1993, audiocassette). This motto resonated well at the end of Pinochet's dictatorship and the beginning of Chile's "democratic transition," with the Native resurgence in Ngulu Mapu and the parallel rise of Mapuche organizations in greater Santiago. Broadcasting an ülkantun connected Indigenous sound to the spirit of resurgence that was permeating Mapuche struggles. In the late 1990s and throughout the 2000s, hegemonic Chilean media had "criminalized" Mapuche land-restoration activism, and the Chilean state had incarcerated a significant number of Mapuche territorial leaders under the label of "terrorists" or similar epithets (Richards 2013, 101–33). Thus, airing an ülkantun that praises a major figure of anticolonial struggle in Ngulu Mapu as a playful and skillful palin player constitutes a powerful counternarrative. How appropriate, then, that the final verses of the ülkantun lyric conclude by praising a proudly victorious Kajfvlikan: "wvle ci antv majmatuiatuay kuzefe kai, / wewgen em ga piawtuai ga kuzefe kai" (Tomorrow the competitor will go around as a presumed one, / "I have been the winner," the competitor will be saying). In times of struggle and resistance, through this chant we hear a voice that both extols and rises toward "tomorrow," a chant of self-affirmation, of a "winning" Mapuche.

VOICES OF A MAPUCHE MOVEMENT

Now it is 8:30 p.m. on Saturday, January 10, 2009. Wixage Anai! *begins to air a special presentation dedicated to the memory of Matías Catrileo, a young Mapuche activist killed a year earlier in Temuco. The special is a recording produced and edited by a member of the* Wixage Anai! *group and available on* CD. *We hear the voices of Margarita Elizabeth Huenchual Millaqueo and José Paillal Huechuqueo reminding listeners that Matías Catrileo was a young Mapuche student who became the victim of a police shooting. While attending one of the public demonstrations that had become so common in Chile, this one calling for the restitution of Mapuche lands, Catrileo was gunned down by a Chilean state trooper. As the special presentation comes to an end, the hosts decry the lack of justice in this specific case, and in more general terms, for all Mapuches in Chile. Eventually Catalina Catrileo and Rayén Navarrete Pailahual, the murdered activist's sister and girlfriend, respectively, join the conversation. They tell stories about Matías Catrileo and describe the events held in Santiago in his honor by both Mapuches and non-Mapuches. Then, in accordance with Mapuche tradition, they speak of dreams in which the image of the young activist becomes manifest.*

....

In devoting airtime to injustice against a Mapuche figure like Matías Catrileo, *Wixage Anai* expanded its objectives from cultivating an Indigenous cultural heritage to promoting political action. The political and cultural rearticulation of Mapuche agency had strengthened considerably since the program first went on the air and the Pinochet dictatorship had ended. However, the 1993 Ley Indígena (Indigenous Law) that subsumed the entity of Indigenous peoples in Chile as *etnias indígenas* (Indigenous ethnicities) conferred penalties of a different kind on minorities. As scholar Diane Haughney points out, in Chile, "State laws concerning Indigenous peoples since the 1800s have generally aimed at the elimination of the protection of Mapuche lands." These policies have attempted to assimilate Mapuches as "individuals" in "Chilean society" and subsequently divest them of their status as "distinctive peoples" (2006, 11). By applying the settler criollo lexicon of *etnias* to folklorically label Native peoples, as if to catch up with the international wave of multiculturalism in the late twentieth century, the Ley Indígena threatened movements of Indigenous pride and self-determination by establishing legal definitions to erase Mapuche cultural identity. This was happening just as the political and cultural rearticulation of Mapuche agency in the 1980s and 1990s was giving shape to a more autonomous Mapuche response to the Chilean settler politics of assimilation and elimination. *Wixage Anai*, with its program on Matías Catrileo, added its voice to this politically charged atmosphere and emerged in active engagement with a process of Mapuche resurgence that was foregrounding a politics of autonomy.

In the years prior to the inauguration of *Wixage Anai*, the radio team members had been involved in other cultural initiatives to foster the movement of Mapuche rearticulation. In the spring of 1991, some of its founding members who were living in Pudahuel organized a Nguillatun. The main ceremony in Mapuche communal life,[16] the Nguillatun gathers the members of a community to express gratitude to the spiritual beings of the earth and the universe (Mapu) as well as to request their help with matters critical to the well-being of the land and the communal life in a territory. It is conducted by a Machi, in close collaboration with the *Ngenpin* (the person who cultivates the art of oratory and preserves and protects ancestral community knowledge) and the longko (the social and political authority in a territory).[17] The local hosts often invite neighboring communities to the Nguillatun, which traditionally takes place every four years. However, many Mapuche communities celebrate it annually, and sometimes it can be convened by a Machi. Since Mapuche musical instruments and chant have major roles throughout a Nguillatun, Chilean scholar José Pérez de Arce accurately defines it as a ceremony characterized by "the sound that prays" ([2007] 2020, 121). The Nguillatun that Mapuche organized in Pudahuel was a true challenge to conceive and carry out given the metropolitan urban environment. Because of its many requirements, many Mapuche inextricably associated the ceremony with a rural setting. Yet in the diasporic context of Santiago, organizing and being involved in a Nguillatun emerged as a genuine spiritual need and desire for cultural reconnection. It also had the underlying political motivation of gathering Mapuche to generate a space of collective self-affirmation.

Speaking about the planning of the 1991 Nguillatun, Ramón Curivil Paillavil recalls that in 1990, after he and his family moved to Pudahuel from Peñalolén, an equally peripheral commune located on the other side of Santiago, he managed to meet a Machi named Augusto Aillapan as well as other Mapuche who, like himself, had migrated to the area. Shortly afterward, in mid-1991, Machi Aillapan invited a group of Mapuche, including members of what became the leadership base of *Wixage Anai*, to organize and participate in a Nguillatun near his home. The Machi's proposal triggered debate among the other Mapuche about whether it was respectful to carry out this type of ceremony in the city. Machi Aillapan had already participated in Nguillatun in Santiago, such as the ones that had taken place in the Commune of Cerro Navia in March 1989, June 1989, and February 1990.[18] Speaking from the authority of his spiritual leadership, he was emphatically affirmative in his response and underscored the Mapuche people's connection to this land.

> El Machi nos dijo, "No pueh peñi. Este territorio era antiguamente de nosotros. Aquí vivió nuestra gente, aquí lucharon. Aquí murió nuestra gente por lo tanto aquí podemos hacer el Nguillatun, en la ciudad, en Santiago, porque además, dijo, la fe es lo que vale." (Curivil Paillavil, interview by author, 2022)

> The Machi told us, "No pueh peñi. This territory used to belong to us. Our people lived here, they fought here. Our people died here. Therefore, we will carry out the Nguillatun in the city, in Santiago, because, moreover, he said what matters is faith.

Based on this sense of "faith," under the leadership of Machi Aillapan, on October 12, 1991, a group of Mapuche carried out a Nguillatun in southern Pudahuel in a public space adjacent to Highway 68 on the way to the Santiago airport. The gathering became the inspiration for future Nguillatun ceremonies in the area. It sparked discussion around two questions. First, how to support the territorial mobilizations of Mapuche communities in Ngulu Mapu? And second, how to have a radio program that could help build connectivity among the Mapuche in Santiago and across Wallmapu? A significant participant in these discussions was the grassroots organization Meli Rewe, which at this time played a role in ensuring the continuity of traditional Mapuche activities such as the Nguillatun over the following years. According to Curivil Paillavil, Meli Rewe was influential in strengthening community ties in several communes in greater Santiago and in helping Mapuche groups to gather for Nguillatun ceremonies and palin.

While Mapuches from the city were asserting their identity, the struggle for land recovery and resistance against the corporate extractivism of forestry companies, hydroelectric projects, and other such ventures had become more active in the southern provinces of Bío-Bío, Arauco, Malleco, and Cautín. When *Wixage Anai* went on the air in June 1993, it transmitted news about community activities in Santiago while also reporting on what was occurring in Ngulu Mapu. Curivil Paillavil maintains that from its inception the radio show expressed its political sympathy for the Mapuche movement:

> El aporte de la radio tiene que ver con la movilización Mapuche porque a partir del 92 prácticamente dos o tres veces al año habían encuentros de organizaciones Mapuche que planificaban una marcha de apoyo a los peñis del sur que estaban en recuperación de territorios y muchos de ellos estaban presos. Entonces todas las marchas que nosotros hacíamos eran apoyadas por la radio. La convocatoria era siempre . . . la radio decía nos vamos a juntar en Metro, y de ahí nos tomamos nomás la calle y marchamos hacia La Moneda. (Interview by author, 2022)

> Radio's contribution had to do with the Mapuche mobilization, because from 1992 on, every two or three years there were gatherings of Mapuche organizations that were planning marches to support the peñis in the south who were working to recover territories, and many were imprisoned. So, all the marches that we did were supported by the radio. The call was always . . . the radio told us to meet at the Metro station and from there we'd just take the street and march toward La Moneda.

For Curivil Paillavil, *Wixage Anai*'s cultural and political alignment was a key component in their Mapuche way of making radio. The team was eager to mesh language, music, and Mapuche cultural values with the struggle to forge spaces in cities like Santiago as well as with the emerging land-recovery movement in Ngulu Mapu. For the Wixage Anai team, the use of radio and sound had to take place in the social and political realm as well as part of broader land relations. Radio space became a site of struggle in which hegemonic enunciative and sonic regimes were re-created and indigenized to articulate Mapuche collective agency engaged in the historical struggle for the cultural and territorial reconstruction of Wallmapu.

The Wixage Anai team, the Jvfken Mapu Center, and Meli Rewe organized Mapuche social gatherings in the Quinta Normal Park to strengthen community rootedness (Cárcamo-Huechante and Paillan Coñoepan 2012, 355–56). By the mid-1990s, the Mapuche at the gatherings in the Parque Quinta Normal were much more open about their Indigenous identity, a fact that resonated with the broader climate of Indigenous resurgence in the capital city as well as the land-recovery struggles in Ngulu Mapu. As a result, the park became a gathering place for other important Mapuche grassroots organizations in greater Santiago, such as Meli Wixan Mapu, the collective Kilapan, and the Coordinadora Mapuche Región Metropolitana.[19] Wixage Anai's promotion of activities in the park contributed to the strengthening of grassroots organizations and positioned the radio program as an active participant in Mapuche community building in Santiago.

Curivil Paillavil asserted that "Meli Rewe and the program never sidelined the political dimension" (interview by author, 2022). Beyond its commitment to share news about the Mapuche movement, such as "the call" to attend a march, *Wixage Anai* itself constituted a form of engagement in actions such as street mobilizations. Mapuche who heard the vocal act of announcing "the call" via microphones or reporting in situ took part in a mobilization as Mapuches, thanks to *Wixage Anai*. "From there, we took over the streets and we marched toward La Moneda," he says, referring to the Mapuche marches in downtown Santiago that gained visible and audible presence in the early 1990s, which usually started in the emblematic Plaza Italia (near the Baquedano subway station) and ended at La Moneda Palace, the main seat of the Chilean government. Curivil Paillavil and countless others thus attest that the radio program was part and a parcel of the Mapuche movement.

A milestone in *Wixage Anai*'s deepening progressive political engagement occurred in 1996, when the team reported on and participated in a march from Santiago to the Parliament building in the city of Valparaíso. Curivil Paillavil notes, "The most important march was when the leaders were in prison, when we walked to Congress" (interview by author, 2022). This march, which was jointly promoted by the Meli Rewe community network, the *Wixage Anai* team, and allied groups, involved around two hundred people walking from Santiago to Valparaíso, approximately ninety-three miles, in support of the Mapuche leaders who had been incarcerated as part of the repressive tactics the Chilean government used against the Temucuicui Lof, a community emblematic of the Mapuche struggle for land recovery in the province of Malleco in Ngulu Mapu.

The Mapuche movement has always incorporated sonorous components in its political marches and rallies. The involvement of radio in activism contributed to this sonic dimension, with its Mapuche musical instruments and slogans in Mapudungun. Street mobilizations included not only speeches, slogans, and chants, amplified by loudspeakers, but also the sounds of Mapuche musical and ceremonial instruments, such as the kultrun, the kull kull, the trutruka, and the pifilka (wind instruments).[20] The native language also inserted an unusual phonetic inflection that upstaged the usual Spanish monolingualism of Chilean street politics. The wide range of sonic resources employed in these mobilizations interfered with the habitual acoustics of public spaces and gave a particular musicality to the Mapuche movement. The incorporation of traditional dance known as the *purrun* added another distinctive aesthetic and cultural flavor to Mapuche marches, transforming the political bodies into a dancing, swirling movement of women on the front line. More than what Western scholars label a "performance," these collective manifestations reflected how Mapuche understand and act on the political and the cultural in a relational sense: Both are part of the continuum of Mapuche being.

This interweaving of the cultural and the political realms resonated like the voices and sounds of the Mapuche movement on *Wixage Anai*'s broadcasts. The echoes of communities struggling to forge linguistic, cultural, and political agency could reverberate on the air. *Wixage Anai* established a counterpoint to the monotone, monolingual rhythms of hegemonic Chilean public space. The producers became adept at employing the communication infrastructure in new and unexpected ways. Writing on the Nasa people of Colombia, media scholar Mario Murillo notes that "Indigenous radio is clearly a response to demands from a community seeking its own space in a media environment seen to be undemocratic." It thus constitutes "a type of public sphere" and "a tool for broader resistance" (2008, 159). *Wixage Anai* created that type of public sphere and used its platform in support of Mapuche community building, resistance to cultural and racial discriminatory treatment, and solidarity with land struggles in historic Mapuche territories. The evocation of Matías Catrileo, the young Mapuche land-rights activist assassinated by a Chilean police officer, was just one of many reverberations of the Mapuche struggle on the radio program's airwaves.

It is now 8:45 on that same Saturday night, January 10, 2009. Richard Curinao, the program's host, starts reading news bulletins and public service announcements for Santiago's Mapuche community. All this information has come to Wixage Anai *via email. While the host reads the news from the station's computer, songs in Mapudungun can be heard in the background. Then the program concludes.*

ACOUSTIC COLONIALISM AND GLOBAL INDIGENOUS RESPONSE

Radio has become a critical form of communication for Indigenous communities, artists, and activists. Recent studies suggest the linkage between radio and Indigenous agency is a global phenomenon.[21] I have demonstrated how radio became a form of interference against colonizing airwaves that affirmed Mapuche resurgence across Wallmapu on both sides of the Andes.[22] Similar processes have taken place in other regions of the continent. Indigenous radiophonic activism has been important on Turtle Island—an ancestral name for what we know now as the United States. Over the years, I have become familiar with the existence of several American Indian and Indigenous radio shows, such as *First Voices Radio* (formerly *First Voices Indigenous Radio*), hosted and produced by Tiokasin Ghosthorse (Lakota).[23] *First Voices Radio* is a one-hour live program now syndicated to seventy radio stations in the United States and Canada, and broadcast from New York City through WBAI 99.5 (Pacifica Radio). The program aims to connect "all the Native Peoples here in Turtle Island (renamed North America by the occupiers)." Kanaka Maoli scholar and activist J. Kēhaulani Kauanui produced and hosted the radio program *Indigenous Politics: From Native New England and Beyond*, which was broadcast from Wesleyan University in Middletown, Connecticut, between 2007 and 2013. Syndicated by stations from different states, this program connected a diverse range of Native American audiences.[24] In California, a team of Mayan and Guatemalan activists produced the radio show *Contacto ancestral*, which defined itself as "a multilingual and multicultural Indigenous program transmitted in Maya K'iche,' Kaqchiquel, Q'anjoba'l, and Spanish." Aired through KPFK 90.7, a Pacifica station in Los Angeles, this radio show became an important medium of linkage for Mayan, Guatemalan, and Central American communities in Southern California.[25] In short, radio has functioned as a connective medium as well as an essential tool in the reconstruction of Indigenous notions of peoplehood, territoriality, and even new diasporic spaces, both on the air and on the earth.

Radio reintroduces sounds, voices, and echoes rooted in ancestral Indigenous traditions. Broadcasting is also an affordable medium for communities that suffer from economic and social marginalization, which is the case for the Mapuche in settler nation-states such as Chile and Argentina. In Chile, the patterns of a history of settlement and colonization continue to inform a mainstream culture awash in prejudice and stereotyping. A 2001–3 study carried out in the Temuco by María E. Merino and Daniel Quilaqueo found that more than 80 percent of 260 non-Mapuche male and female interviewees engaged in practices of "ethnic discrimination." In their answers, the interviewees used labels for the Mapuche people such as "primitive," "uncivilized," "ignorant," "lazy", and "uneducated," among other racially stigmatizing terms (Merino and Quilaqueo 2003, 109). The converse of such negative stereotyping is also present in the paternalistic discourse of the middle and upper classes, New Age groups, museums, and mass-media outlets linked to private capital or the Chilean state that extol the Mapuche people as bearers of "ancient spirituality and wisdom," thus erasing the political and historical dimensions of the Mapuche past and present. Finally, the proselytizing activities of powerful mainstream Catholic and evangelical media networks indirectly reinforce stereotyping with the religious booklets, pamphlets, radio programs, press, and door-to-door missionaries they use to permeate Mapuche communities and influence contemporary religious and cultural practices.

Voicing their linguistic, cultural, and political differences through Western technology like radio, the press, and the internet has become an imperative for activists invested in enabling Mapuche people to fashion their own styles and politics of self-representation. During the 1990s, media activism emerged as part of a collective struggle for voice in a mediascape still mediated by the specter of seventeen years of dictatorship and a "democratic transition" trapped within the local political and entrepreneurial elites' agenda, oriented toward the stabilization of market capitalism in Chile. Within this hegemonic neoliberal context, the official celebrations in 1992 of the Fifth Centennial of Columbus's arrival to the hemisphere provoked a countercultural mobilization among Mapuche communities that mirrored the political and cultural path First Peoples undertook throughout the Americas.

The Mapuche use of radio is part of a broader process of activism in the late twentieth and early twenty-first centuries, when the appropriation of outlets and self-representation within the media emerged as strategic components of the cultural and political struggles of Indigenous peoples. The global emergence of "Indigenous media" leads scholars Pamela Wilson and Michelle Stewart to conclude that "control of media representation and of cultural self-definition asserts and signifies cultural and political sovereignty itself" (2008, 5).

Recent studies confirm the worldwide media corporatization's impact in Chile. Radiophonic hegemony is still a substantive force in the deployment of acoustic colonialism. Ironically, though, it gives a decolonizing force to the radio and media initiatives that emerge from local and Indigenous communities. With their divergent sounds and voices, they manage to convey aesthetic, cultural, and political influence, accessible by simply tuning in to a different frequency on the radio dial and becoming immersed in alternative acoustic ecologies. The ethos of Mapudungun invites one to envision sound in conjunction with the land, the environment, and the universe. Meanwhile, the omnipresence of radio complexes, intoning the scripts of corporate capital, embody the monotony of settler colonial society's voice. The challenge then is to create and maintain polyphonic environments. Indigenous media could provide the necessary framework to foster decolonized relations between sound technologies and Native peoples, nations, and territories.

Wixage Anai's experience illustrates the urgent need to keep initiatives of this type in the ether and to challenge the dominance of colonial mediation in the mediascape. During the 1990s and early 2000s, the radio program had a transborder impact. In addition to its main emission from Santiago, *Wixage Anai* was retransmitted on radio stations in

Ngulu Mapu. It reached communities in provincial towns and villages, aurally stimulating a sense of peoplehood as part of Ngulu Mapu and, more broadly, of Wallmapu. *Wixage Anai* speaks to how the aural and sonic connectivity of radio can, effectively and affectively, contribute to the historical process of reconstructing Indigenous autonomies in Wallmapu and around the globe.

Wixage Anai demonstrates how an Indigenous radio program interfered with acoustic colonialism, inciting the political and cultural desire to transform our aural and sonic environments and practice nütram and allkütun in the public realm. From a Mapuche audio-territorial angle, we must not give up on projects that aim to reconstruct and recreate the acoustic ecology of Wallmapu. The *Wixage Anai* producers' radiophonic intervention embodied a Mapuche way of fulfilling a cultural and political need. It proclaimed a desire for a more heterogeneous and democratic public sensorium. *Wixage Anai* made a critical difference in a Chilean society that was shaped by the colonizing "noises" of corporate media. Its story exemplifies how to exercise an autonomous Indigenous politics of communications to build public audibility for the Mapuche, to forge agency by using radio as a medium of connectivity in times of dispersion, and to assert Indigenous voices and sounds on the global airwaves.

FIVE

......

Enduring Listening and Sounds: The Contemporary Musics of Ngulu Mapu

During my childhood in rural Tralcao, from Monday to Friday, I would wake up early and run some one thousand feet from my grandmother María Clara Lefno Huechante's house to my elementary school. Later, ambling home more casually, I would hear Mexican corridos and rancheras playing from my family's Panasonic radio. My grandmother, along with my two uncles at times, would have turned the radio on to Radio Camilo Henríquez or Radio Baquedano—stations from Valdivia, some eighteen miles south of Tralcao—to hear the popular Mexican music they featured that never failed to animate her after her afternoon nap. Tuning in to the Mexican music radio shows commonly known as "los Mexicanos," was a daily habit among the Mapuche-Williche families who were the majority of Tralcao's inhabitants. The programs also functioned as community bulletin boards, and between songs they enjoyed, my Tralcao neighbors also looked forward to hearing the broadcaster, with a Mexican "charro" accent, read messages to family and friends scattered between the city and the country that listeners sent to the station.[1]

Once I was back from school in my family's home in the Füta Willi Mapu (the Great Land of the South), I would tune in to

Radio Concordia, from the town of La Union, or Radio La Voz de la Costa, from the more distant city of Osorno, for other sources of popular Mexican music. By the end of the afternoon, with my grandmother's permission, I switched the radio to rock and ballads in English. Though I didn't understand them, they constituted an intriguing childlike escape from the usual predominance of corridos and rancheras.

At nightfall I often heard different sounds after my grandmother sent me out wandering through the small family farms to find our flock of sheep and bring them home to their pen. Few fences separated farms in Tralcao, where traditional Mapuche-Williche lifeways prevailed and passing near the house of elder Reynaldo Martin Pangui while looking for sheep, I remember hearing Reynaldo seeming to talk with the traru (birds) that sang high in the large cherry trees around his home, where his kitchen's hearth was always burning. Many times, I would hear him singing. Always a curious child, I would stop to listen, even without fully understanding what he sung.

Years later, on a summer evening on December 24, 2023, under the shadow of a cherry tree on the same lands, Reynaldo's son, Manuel Martin Mora, confirmed that his father often kept up a low chant in Mapudungun to keep the traru away while he burned a handful of sheep wool in the hearth. He also described the more eventful auditory moments in his childhood home that occurred when Reynaldo gathered with other elders of the Pangui kinship to parlamentar *and* romancear, *to talk and chant in Mapudungun. Elder Reynaldo Martin Pangui, a Lafkenche (Mapuche from the Coast) who had moved from Pichikullin, near the rural, coastal area of the Commune of San José de la Mariquina to Tralcao when he was seven years old, passed away in 1972. According to his son, Reynaldo's skill as an ülkantuchefe (chanter) was widely known throughout Ngulu Mapu. A Mapuche medicinal and spiritual authority, or Machi, named Antiman from Temuco, a city some ninety miles north to Tralcao had even invited him to travel to Puel Mapu (what is today called "southern Argentina") to perform in or lead ceremonies with chant. This invitation from beyond the Mapuche-Williche territory bears witness to Elder Reynaldo's renown in the art of ülkantun. By the 1960s and early 1970s Elder Reynaldo Martin Pangui was one of the few practitioners of ülkantun and speakers of the native language to survive the process of Chileanization and Christianization in Tralcao and his ülkantun embodied a rare legacy of Mapuche culture.*

The voice of Reynaldo Martin Pangui still resonates in my aural memory, testifying to Mapuche endurance. My attentive listening to his singing as a child was a way to engage with Mapuche music in a colonized environment in which our ears were fed a steady diet of non-Native sounds, both on our own radios and at festive community events. From morning to night, the radio played romantic ballads in Spanish, and, less often, in English or French. We heard Chilean cuecas, and Mexican corridos and rancheras, Colombian cumbias, rock in English and sometimes rock in Spanish. Some stations played European classical music as well. We began the week at my rural elementary school by having to sing the National Anthem of the Chilean settler nation-state. We also chanted Catholic hymns and prayers in the small chapel on Sundays. Alongside these sonic layers of a colonized Mapuche-Williche environment, those other fleeting moments of listening to Reynaldo Martin Mora's slow, whispering ülkantun left an indelible trace in my memory, the reverberations of the abiding indigeneity of Ngulu Mapu acoustics. The other types of music never resounded in the same way.

....

This chapter focuses on how Mapuche listening and vocal performances have been manifested across different genres. The recontextualization of non-Mapuche musical expressions and native forms of music demonstrate a rich variety of auditory and musical practices. I contend that Ngulu Mapu constitutes an "acoustic territory" (LaBelle 2010) in which ears and voices of the Mapuche stage complex ways of engaging with music. I forcefully argue in this chapter that the widespread impact of radio, which definitely took off in the late 1960s and 1970s, was a key factor in the popularization of rancheras and corridos in Ngulu Mapu / Chile. The Mexican films that came to Chile in the late 1930s and expanded in influence in the following decades, paralleled by the relative accessibility of victrolas, also played some role in this process.[2] Then, in the chapter, I discuss the more recent phenomenon of rap as a genre that makes inroads among new generations of Mapuche in the politically turbulent 2000s. I conclude by offering an overview of the continuity of ül, a type of vocal and instrumental musical genre that has a long-standing tradition in Mapuche life.

I highlight how these forms of listening and vocalization engage the resonant capacities of the heart and mind. This premise derives from a cardinal principle from Mapuche Ngen (being Mapuche): when we listen to music, we feel and also think. Consequently, the Mapuche are predisposed to endow sound with a sentient and thinking dimension culturally and historically linked to the Mapu—the land, the territory, the universe. I put this Mapuche relational approach in dialogue with recent scholarly elaborations on the generative roles of listening, voice, and music.

I adopt musicologist Ana María Ochoa Gautier's premise that listening plays a critical role "in the constitution of acoustic ontologies and knowledges" in contexts of colonization (2014, 34–35). Musical listening is a generative force in the modes of being. The continuous metamorphosis of sounds in local contexts is not passive. Moreover, Licia Fiol-Matta's formulations on how a "thinking voice" takes shape in the performances of popular female singers in Puerto Rico has helped me develop my approach to the role of Mapuche vocalities in the cultivation and re-creation of ül and rap. Fiol-Matta questions whether the voices of musical artists are mere acts of "spontaneous execution." She regards the voice as a "producer of thought" (2017, 7–8). In a Mapuche context, producing and listening to vocal music require generative ears and voices, ontologically, epistemologically, and historically. In this sense, musical listening, vocalization, and sonic creativity enable us, the Mapuche, to forge modes of acoustic habitation and Indigenous endurance.

The concept of music that streamed from the Chilean radio industry often was antithetical to the musical listening and vocal practices of the Mapuche. South of my own region, the main radio station was Radio SAGO, an offshoot of the Agricultural and Livestock Society of Osorno (hence the acronym SAGO in Spanish), the major association of the region's colonial elite of *terratenientes* (wealthy landowners).[3] Since its founding in 1939, the core of Radio SAGO, like many other radio stations, was invested in a Chilean settler vision, with a racialized way of relating to the Mapuche. Radio SAGO is significant here, though. Juan Atanacio Meulen Tranayado (Juan Meulen from now on), who became a notable Mapuche-Williche figure in radio communications and culture in the late 1960s, remembers that he often heard Radio SAGO officials make racist, anti-Indigenous statements, including a representative of the radio station once saying, "Los Mapuche no venían a cantar sino a balar a la radio" (The Mapuche didn't come to sing but to bleat on the radio) (interview by author, 2015). To this man, the Mapuche were incapable of mastering the art of music (*no venían a cantar*) and were only able to bleat (*balar*). According to the Real Academia de la Lengua Española's dictionary, *balar* refers to emit *balidos* (bleats), that is, the "voice of the ram, the lamb, the sheep, the goat, the fallow deer and the deer" (2001, 277). The Radio SAGO official's derogatory comment on the vocal performances of the Mapuche exemplifies the deeply embedded Creole tendency to hear the sounds of "the native" as primitive, as both infrahuman and infra-animal. Unsurprisingly, this parallels stereotypical judgments I have already unveiled in previous chapters of this book, witness the linking of Mapuche collective voicing to "chivateo" in nineteenth-century Creole literature, in chapter 1, or, in chapter 2, the vocal contortions of the Mapuche in the strident vocality of Indio Pije in cumbia songs and comic radio shows of the 1960s and 1970s.

The acoustic disfigurement of Mapuche voices mirrors logics of racialization that aims to derogate the very humanity of the non-White selves. As pointed out by Black studies scholar Alexander G. Weheliye, racialization deploys in society "as a conglomerate of sociopolitical relations that discipline humanity into full humans, not-quite humans, and nonhumans" (2014, 3). In what Juan Meulen describes, Indigenous subjects are racialized as entities who cannot sing but simply "bleat," a projection that derives from stereotypes associating the pastoral domain with the rustic and guttural, the infrahuman and the infra-animal, in contrast to the supposed sophistication and urbanity of Western music. A racial and geographical hierarchy emerges from this anthropocentric, White urban order. The stigmatization of Indigenous voices entails a derogatory perception of the animal and the rural that relegates them to an inferior aesthetic, cultural, and ontological status. Within this adverse milieu, the Mapuche has been able, however, to stage a heterogeneous idea and practice of music through modes of listening, vocalization, and instrumentation that defies the narrow colonial ear and mindset of the dominant society.

"LOS MEXICANOS" ON THE AIR!

Mapuche-Williche poet and writer Graciela Huinao's 2010 novel *Desde el fogón de una casa de putas Williche* (From the Hearth of a Williche Whores' House) plunges readers into the environment and stories of a brothel staffed and frequented by Mapuche-Williche in Rawe Bajo, a working-class barrio alongside the Rawe River in the city of Osorno.[4] In a striking way, already by the beginning of the novel, we as readers/listeners become promptly immersed in the musical and affective environment of Rawe Bajo:

> Desde uno de estos sacrificados hogares, que tuvo el coraje de ahuyentar las penas, por una ventana se asoma una colorida "ranchera," canto popular de México que el Mapuche adoptó para enamorar el corazón a orillas de una fiesta de campo y ciudad.
>
> No atinan de dónde les apuntaron el dardo, pero la inyección musical en los ancianos fue a las venas. Los músculos centenarios levemente empezaron a moverse en sus rostros, la picardía de una sonrisa empieza a jugar entre los mustios labios y sus engarrotados dedos empezaron a soltarse de los bastones . . . La música suelta los recuerdos, éstos avanzan y en medio de la calle se abrazan al ritmo de una ranchera. (Huinao 2003, 22)

> From one of these self-sacrificing homes, one that had the courage to frighten off sorrow, through a window appears a colorful "ranchera"; a popular song from Mexico that the Mapuche adopted to get the heart to fall in love at the edge of a party in the country and the city.
>
> Despite not knowing where the shot of sound came from, the musical injection went straight to the veins of the male elders. Ever so slightly, the aged muscles of their faces began moving, a cunning smile starts to play between their withered lips and their stiffened-up fingers began to drop their canes. . . . Music releases their memories, they walk forward; and, in the middle of the street, they embrace each other to the rhythm of a ranchera song.

This passage is a record of a Mapuche aural environment in which rancheras often predominate. Huinao's synesthetic figuration of music places readers into a visual and sonic event: "A colorful 'ranchera' appears." Sound acquires visuality. Readers are invited to see, hear, and feel music on multiple levels that surpass the oral and auditory domain of the ear and engage a broader sense of "aurality" (Ochoa Gautier 2014), that is, to experience the blending of the communal, affective, and carnal. As expressed so poetically by Huinao, the "popular song of Mexico" affects "the veins," "the muscles," "the memories," and even the physical gestures and movements of the Mapuche-Williche. Feelings, images, sounds, country-city territories, bodies, and rhythms intertwine "to the rhythm of a ranchera song." Huinao's novel emphasizes how Mapuche people adopted Mexican popular music "to get the heart to fall in love at the edge of a party in the country and the city." Her poetic narrative offers us an eloquent register of a significant collective aural experience: the Mapuche musical taste for rancheras and corridos.

These two types of Mexican music became popular among Mapuche listeners thanks to the Victrola and records and, then in the following decade, the "Mexicanos" shows that radio broadcasts made it possible to hear in homes, workplaces, and parties in the countryside and in the cities where Mapuche had migrated.[5] Although the literature on Mexican popular music tends to draw genre distinctions between rancheras and corridos, over time, both musical genres coalesced as one single great musical chord in the ears and hearts of the Mapuche. In his influential book *El corrido mexicano*, originally published in 1954, Mexican musicologist Vicente Mendoza states, "El corrido es un género épico-lírico-narrativo, en cuartetas de rima variable, ya asonante o consonante en los versos pares, forma literaria sobre la que se apoya una frase musical compuesta generalmente de cuatro miembros, que relata aquellos sucesos que hieren poderosamente la sensibilidad de las multitudes" (The corrido is an epic-lyric-narrative genre, in quartets of varied rhyme, either assonant or consonant in the even verses, a literary form in which four members generally perform a musical phrase that tells of events that powerfully affect the sentiments of the crowd) ([1954] 1976, ix). In sum, as a genre linked to the tradition of the romance in Spanish, the corrido tells a story.

Magdalena Altamirano characterizes the corrido "as a narrative song," one that it is almost always composed "by a series of eight-syllable quatrains, with their own rhyme in the even verses" (Altamirano 2007, 261). Through its very name, the ranchera is associated to the life of Mexican villages, *ranchos*. Rancheras are often couplets more varied in meter and measure than corridos. In any case, both Mendoza and subsequent scholars affirm the epic and narrative quality that organizes the lyrics, structure, cadence, and rhythm of corridos. In contrast, the sentimental is more pronounced in rancheras. Dutch scholar Claes af Geijerstam distinguishes between the two genres by stating that "the main difference between songs and corridos is that the song has a lyrical, often sentimental quality in the text and melody, while the corrido has an epic or narrative quality" (1976, 59). Despite their differences, however conceived, both rancheras and corridos have been part of the same Mapuche aurality, thanks to radio shows devoted to Mexican popular music.

The first Mapuche to create and host a Mexican music radio program in Ngulu Mapu was Juan Meulen (1945–2025). Born in Quilquilco, a historically Williche community in Chaura Kawin, one hundred seventy-four miles east of Osorno, he has been able to recount the shifting winds of the Mapuche musical landscape as he experienced them himself, as both child and adult. As Meulen recalls, the traditional instruments of Mapuche musical culture, such as the kültrun and trutruka, were used only rarely in Williche communities, reflecting the effects of colonization in the region. Few families had a trutruka at home, and those who did played it only occasionally. As a child growing up in those times without radio or television, Meulen was exposed largely to non-Indigenous music, usually experienced through live performance at community events featuring the participation of local performers of various non-Indigenous musical genres, such as the Chilean cueca, *pasacalle* (a musical expression originating from of the Chiloé Island) and, of course, rancheras. Due to the perquisites of these musical genres, the guitar and the accordion became the instruments most embedded in the aural environment of Chaura Kawin, as was the case also in a large part of the Füta Willi Mapu during the second half of the twentieth century. From those years, Juan Meulen remembers,

> En todas las festividades que se hacían se tocaba música mexicana. Claro, después salieron las victrolas; y la gente compila esos discos con música ranchera. Yo crecí escuchando música ranchera. Después, cuando recién llega la radio, por ahí por los años . . . serían casi el 60', por ahí, llegan los primeros receptores a la comunidad. (Interview by author, 2015)

> In all our festivities, Mexican music was played. Of course, the victrolas arrived and people started collecting ranchera records. I grew up listening to ranchera music. Later, when radio arrived, more or less in the 1960s, the first radio receivers arrived in the community.

In live performance and records on a Victrola, Meulen confirms that what stands out is the aural traction that rancheras and corridos gained among the Mapuche. In his nostalgic account of his experience and that of his Mapuche-Williche generation as captivated radio listeners in Quilquilco in the early 1960s, he recounts,

> Eran muy pocos los que tenían un receptor de radio. Entonces concurríamos nosotros los jóvenes en ese tiempo donde el amigo, el que tenía una radio, íbamos a escuchar algunas veces la radio. Y escuchábamos ahí la música mexicana; en ese tiempo, la radio, cuanto era . . . , de Santiago, que tenía un programa. Las radios venían en onda larga y onda corta; y en onda corta se escuchaban las radios santiaguinas acá, y ahí, a las 3 de la tarde, había un programa de música ranchera. (Interview by author, 2015)

> Few people had radio receivers. So, we would go to a friend's house, the one who owned a radio. And we listened to Mexican music; at that time, the radio station, I try to remember which one . . . , from Santiago, it had a program. Radios were on broad waves and short waves; and on short waves, we managed here to tune in the stations from Santiago, and there, by 3 p.m., there was a radio show of ranchera music.

As Meulen notes, in rural areas like Quilquilco, it was difficult to connect to the radiophonic waves or access radio transmissions. Few homes had radios, so listening to Mexican popular music around a radio became a gregarious social experience for Mapuche youth of Meulen's generation. Trying to tune into a sought-after Mexican music program broadcast from a Santiago station and only accessible on shortwave, Meulen and his friends visited the compadre who had a radio at home. The rancheras program that Meulen mentions was broadcast by the Santiago-based Radio Yungay, one of the oldest radio stations in the country.

The 1960s was the decade of the "transitorización" of radio culture in Chile, when portable radio receivers spread nationwide, according to radio historian and longtime radio broadcaster Luis Gamboa (interview by author, 2024).[6] In the southern provinces, local radio stations that had been founded in the 1950s consolidated, and some new stations went on the air. By the mid- to late 1960s some of them were airing shows similar to the one Juan Meulen and his friends used to hear. Radio Eleuterio Ramírez, from Osorno, broadcast the program "Música y Canciones de México"; and Radio Concordia, from the small town of La Union, also played selections of Mexican popular music.[7] As several radio stations in Ngulu Mapu broadcast rancheras and corridos, these genres started achieving unmatched popularity in Mapuche homes. Toward the end of the 1960s and early 1970s, besides the previously mentioned stations, a significant number of stations already had their own Mexican popular music programs within the Füta Willi Mapu; among them, Radio Pudeto de Ancud, on the Island of Chiloé; Radio Vicente Pérez Rosales from the coastal city of Puerto Montt; Radio La Voz de la Costa from Osorno; and Radio Baquedano and Radio Camilo Henríquez, from Valdivia. Farther north in Ngulu Mapu, the sounds of rancheras and corridos were also aired by numerous radio stations, such as, to name a few, Radio Aníbal Pinto from Lautaro, in the Province of Cautín; and Radio Los Confines from Angol, in the Province of Malleco; and Radio El Carbón from Lota, in the Province of Arauco. In the everyday lexicon of the region, these programs, no matter the station that aired them, became known collectively as "los Mexicanos."

By the end of the 1960s, Juan Meulen was employed by the Rahue Mission, which owned Radio La Voz de la Costa. The radio station had been founded in 1968 by the Rahue Mission, a group of mainly Dutch priests who settled in Chaura Kawin to expand Christian evangelization of Williche and Lafkenche communities. They established what they called Santa Clara Radiophonic Schools in which they mixed their Christian evangelizing with radio lessons in mathematics, Spanish, agriculture, cooking, law, commerce, and basic commodity prices. Whether or not all listeners were influenced by the Rahue Mission's evangelizing religious, ideological, and cultural colonialist broadcast goals, some in the community, like the young Juan Meulen, who did not know how to write or read and lived far away from any brick-and-mortar school, benefited from the radio 's educational initiatives in writing and reading. This enabled them to acquire the tools of Western literacy, such as writing and reading in Spanish, that were essential to function in settler colonial Chile.

Meulen's experience of a radio "school" began his immersion in techniques of radio communication and paved the way for his aspiration to create his own radio show that tapped into the popularity of "Los Mexicanos." Radio Eleuterio Ramírez, which Meulen called "the radio of [Mapuche] communities" (Meulen Tranayado, interview by author, 2015), had been a pioneer through its afternoon show "Discoteca Regional." However, Radio La Voz de la Costa had also reached a significant audience in Mapuche and non-Indigenous campesino communities in

Chaura Kawin. Although he wanted to broadcast a Mexican music program on this station, he had problems doing so:

> La música mexicana se escuchaba y . . . yo planteaba siempre. . . . Entonces lo que me decía el director de la época, de esos años, el Padre Manuel Sánchez, "No, que la música mexicana habla de puras muertes, de borracheras, que lo engañó, en fin . . . ; y esto es una radioescuela." Entonces yo siempre insistía, insistía. Yo decía: "Pero ¡no todas las canciones hablan de muertes y de borracheras!" (Meulen, interview by author, 2015)

> Mexican music was heard quite a bit and . . . I always highlighted it. . . . Then what the station's director at the time, in those years, Father Manuel Sanchez, "No, that Mexican music speaks only of deaths, drunkenness, love deceptions, anyway . . . ; and this a radioschool." So I always insisted, insisted. I said to him, "But not all Mexican songs speak of deaths and drunkenness!"

The Catholic priest was not alone in stigmatizing Mexican music by associating it with alcohol consumption, violence, and drama and pontificating against it from a Christian perspective. To him, it was just cantina music. By countering the approach of the Catholic priest, Meulen offers a different perspective on the variety of the corrido and ranchera tradition. In the introduction to his anthology *Lírica narrativa de México: El corrido* (1964), Vicente Mendoza transcribes the lyrics of diverse corpus of corridos to prove that the genre covers varied registers: "Revolutionary corridos" address the Mexican Revolution; "lyrical corridos" iconize revolutionary heroes and the idealism of struggle. "Bandit corridos" and "prison runs" sing of "patricides, curses and doom." Others express "religious, biblical, and moral themes" or praise "the merits and beauties of cities." Corridos discuss a variety of "events that shake the daily lives" of the people in Mexican rural environments (Mendoza 1964, 31–41); subsequently, the reductive association of corridos and rancheras with cantinas appears simply to be an ideological prejudice.

Despite the initial opposition of the Radio La Voz de La Costa's director, "Cancionero Mexicano" (Mexican Songbook), Meulen's radio show, debuted on the air in the middle of 1971, with Meulen serving as announcer. Using a record player, the musical repertoire of his program included rancheras and corridos by classic Mexican performers such as José Alfredo Jiménez, Antonio Aguilar, Francisco "El Charro" Avitia, among others, as well as the "Mexican music" by artists like Guadalupe del Carmen and the Hermanos Bustos, who would be pioneers in cultivating and promoting Mexican popular music in Chile. To expand the program's musical offerings, Meulen enlisted the aid of friends who traveled to Mexico and brought back records while he bought whatever new materials were available at shops in Santiago. At first "Cancionero Mexicano" was broadcast for only fifteen minutes per day, but it soon doubled its airtime to thirty minutes in response to the demands of its rapidly increasing audience.

In its inaugural months, the program received an average of three thousand letters per week, with various greetings and messages listeners wanted the station to read to the on-air audience, a trend that continued over the years. These messages enabled "Cancionero Mexicano" like most of *los Mexicanos* in Ngulu Mapu, to serve as a kind of community message board, transmitting communications from the audience to families and communities in its listening area. These messages were usually read by the broadcaster between songs, though given the quantity of letters sent to the station, the announcer often had to summarize. Meulen recalls with enthusiasm the broadcast of June 24, 1973, the date of the feast of Saint Juan, when they received an avalanche of greetings that kept them on air from 3:00 p.m. to 11:00 p.m.

The program's popularity was also responsible for diversifying the repertoire of the local music industry and enhancing popular enthusiasm for *música ranchera*. Soloists and bands of corrido and rancheras emerged in Santiago and the southern provinces of Chile. Los Reales del Valle, Los Luceros del Valle and Los Tejanos de América appeared in central Chile. In the Chaura Kawin area, local performers also emerged, among them Los Manantiales, Los Diamantes Sureños, Los Praderos, and Los Regionales. With these new voices and sounds at their disposal, los Mexicanos expanded their musical programming.

With Meulen as on-air host, "Cancionero mexicano" at Radio La Voz de la Costa became a popular radio show in Füta Willi Mapu. Meulen produced and directed the program between 1971 and 1990. Throughout the life of the program, as he could tell from the surnames on his listeners' messages, his audience was mostly Mapuche. Nonetheless, Meulen conducted his show in Spanish. The reason for this, he says, is that "discrimination was high. No one wanted to speak the [Indigenous]

FIGURE 5.1 · Outdoor fundraising benefit concert for repair of Radio La Voz de la Costa's antenna, Osorno, Chile, May 1983. On stage, emcee Juan Atanacio Meulen Tranayado (*left*), with a local group of corridos and rancheras. Courtesy of Juan Atanacio Meulen Tranayado.

language." In this setting of linguistic and cultural loss, non-Indigenous musical genres such as rancheras and corridos became an aural refuge. Rancheras, as the novelist Graciela Huinao notes, touched the "heart" of those Williches who imagine themselves "at the edge of a country or city party."

The emotional effect of rancheras has been widely acknowledged in scholarly literature. In *El folklore y la música mexicana* (1928), Rubén Campos highlights the genre's sentimental imprint. He comments, "In the song, the musician has nothing more than sentimentality concentrated in a brief and clear form" (Campos 1928, 80). In another study, Geijerstam notes, "The performance of rancheras exhibits an inimitable sentimentality, even tearfulness, which seems to be characteristically Mexican" (1976, 125). More recently, however, anthropologist Olga Nájera-Ramírez has questioned Geijerstam's designation of "sentimentality" as a "Mexican" peculiarity. In her discursively situated analysis, she argues that rancheras revolve in complex ways "around fervent feelings toward particular people and places," performed through "intense expression of emotions" (Nájera-Ramírez 2007, 457).[8] The ranchera's expressiveness draws on the *voz lastimera* (a plaintive voice). The inclusion of this long, wavering, high-pitched vocal sound, or sort of hoot (*ululātus*), which includes cries full of emotion, heightens the sentimental trait of rancheras and provides it more subtly than Geijerstam acknowledges.

This sentimentality also runs through the lyrics, the vocals, and the very reception of corridos. Both genres cohabit, overlap, and fuse in the aurality of Ngulu Mapu. Geijerstam believes that the style of lyrical, vocal, and performative sentimentality present in both genres blurs the dividing lines between them. The classic corrido "Juan Charrasqueado," for example, composed in Mexico around 1945 and performed by Jorge Negrete in the film of the same name, has a markedly sentimental touch. Like many Mexican popular songs at the time, it emphasizes "manliness"—machismo, to be clear—by praising the womanizing macho Juan Charrasqueado.

"Juan Charrasqueado" was one of the most popular corridos in Mapuche and peasant communities from the 1960s to the 1980s. Honoring the epic-narrative character of every corrido, this song tells a story with a strong emotional charge:

> Voy a cantarles un corrido muy mentado,
> lo que ha pasado allá en la hacienda de la flor,
> la triste historia de un ranchero enamorado
> que fue borracho parrandero y jugador.
> Juan se llamaba y lo apodaban "Charrasqueado,"
> era valiente y arriesgado en el amor,
> a las mujeres más bonitas se llevaba,
> de aquellos campos no quedaba ni una flor.
> Un día domingo que se andaba emborrachando
> a la cantina le corrieron a avisar:
> "Cuídate, Juan, que ya por ahí te andan buscando
> son muchos hombres no te vayan a matar." (Negrete 2002, 65)
>
> I shall sing a very well-known ballad
> about what has happened in the hacienda La Flor
> It is the sad story about a rancher who was a lover,
> a drunkard, a party lover, and a gambler.
> Juan was his name, but he was nicknamed "Scarface."
> He was bold and daring in matters of love.
> He ran away with the most beautiful women,
> so that there were hardly any flowers left in those parts.

One Sunday while he was getting drunk,
someone ran into the saloon and warned him:
"Be careful Juan, many men are looking for you.
I hope you don't get killed."

The story goes from the epic persona of "Juan" as "bold and daring in love" to his tragic fate of a death foretold. The middle stanzas detail his murder: "'Estoy borracho,' les gritaba; 'y soy buen gallo' / Cuando una bala atravesó su corazón" ("I'm drunk," he yelled at them; "and I'm very macho" / When a bullet went through his heart). If melodrama is often associated with the ranchera, the story of this corrido adopts a similar verbal and thematic expressiveness, as evidenced in its two final stanzas:

En una choza muy humilde llora un niño.
y las mujeres se aconsejan y se van.
Sólo su madre lo recuerda con cariño.
mirando al cielo llora y reza por su Juan.
Aquí termino de cantar este corrido.
de Juan ranchero charrasqueado y burlador.
que se creyó, de las mujeres consentido
y fue borracho parrandero y jugador. (Negrete 2002, 65)

In a humble shack a child cries
while the women murmur to each other and leave.
His mother comforts him with love
and looks to the sky, crying praying for her Juan.
I finish here singing this ballad,
about Juan the scarfaced rancher and seducer of women.
He thought that women loved him dearly
and was a drunkard, a party lover, and a gambler.

Interpreters of "Juan Charrasqueado" vocally inflect the penultimate stanza, employing the *voz lastimera*, an affected and plaintive voice that slows the cadence. This can be heard in different interpretations across time, such as the original by Jorge Negrete in the 1940s and later versions by singers such as Charro Avitia or Chavela Vargas. This corrido exemplifies the sentimentalism of the corrido tradition. In the early 1970s, I remember listening to the emotional force of corridos that alternated with rancheras in the musical programming of "los Mexicanos." Rancheras and corridos has resounded in Ngulu Mapu for more than half a century with a sentimentality that oscillates between euphoria, nostalgia, and the melancholia of a people immersed in loss and fragmentation. This split of Mapuche listening reflects well what Tanana Athabascan scholar Dian Million has described as the dimension of "emotional colonialism" (2013, 46).

At another level, corridos evince a patriarchal and heteronormative logic, and their machismo has been widely examined in scholarly literature (Paredes 1963; Chávez 2017).[9] Yet the audience of the Mexicanos in homes, workplaces, and community settings in Ngulu Mapu always has included women and a heterogeneous social body that challenges the monopoly of heteronormative masculine modes of listening among the Mapuche. Women's voices in ranchera or corrido interpretations further reshape this sonic environment. "Juan Charrasqueado" acquires gender inflection through nonconventional—"queer"—singers such as the influential Mexico-based Costa Rican singer Chavela Vargas.[10] Her singing had as much influence as that of her male peers, like Antonio Aguilar and similar charro singers of the second half of the twentieth century. Considering the dominant patterns of ranchera music as a genre "predominantly reflective of male experience," scholar Ana Alonso-Minutti remarks, "Chavela queered these conventions with her 'masculine,' raspy voice" (2020, 50). In the context of Ngulu Mapu, this butch, lesbian, and to some extent transgender voice resembles the almost masculine-sounding voices of campesino women. Indeed, Vargas's interpretation of "Juan Charrasqueado" differs from conventional vocalizations by charro singers and troubles the man/woman binary that the song's lyrics foster. Her treatment of the two last stanzas accentuates the emotional and vocal effects of the *voz lastimera*, transmuting it through her heavy, gravelly voice and giving a different sonic aura to "las mujeres" in the lyrics that blurs gender lines in the aural space.

Chilean singer Guadalupe del Carmen (1931–87), a precursor in the promotion and staging of charro music in South America,[11] gave a quite different spin to "Juan Charrasqueado" during the 1960s and 1970s. She projected an emphatically feminized version of the song by employing a high-pitched voice and festive, bouncy tone instead of the traditional soulful *gritos* and mariachi ensembles. Another important woman was Lupita Suárez, who at Radio Yungay conducted *Mexico es así*, a pioneering show devoted to Mexican popular music in Chile in the 1940s. This weaving of women's listening and vocalities shaped circulation and reception of Mexican popular musics in the acoustic and affective territory of Ngulu Mapu, and complicates the patriarchal, heteronormative script of the ranchera and corrido.[12]

Other broad sociocultural and historical motifs also drive the Mapuche identification with rancheras and the corridos across genders and generations. The Chilean literary critic Lucía Guerra has highlighted how corridos, especially those "focused on dispossession and injustice, violence, and heroes who fight to vindicate the victims of misery and inequality," have become appealing for the Mapuche. Guerra infers that this "thematic content" generates a "preference for corridos" among Chilean audiences, especially those with "a strong identification with the history and marginalization of the Mapuche people" (Guerra 2014, 125). However, Guerra's view is extremely limited, considering that the rancheras and corridos popularized in Ngulu Mapu included a very wide range of themes. Most of "Los Mexicanos" did not play the "revolutionary corridos" of the 1920s and 1930s; and even fewer connected to socially engaged artists, such as, for example, Amparo Ochoa, who cultivated corridos and rancheras within the New Latin American Song of the mid-1960s and early 1970s. Thus, the Mexican popular music programs across decades in Ngulu Mapu did not usually touch on themes like "dispossession," "injustice," "misery," or "inequality," as suggested by Guerra. Rather, the contents of the kind of Mexicanos that most of us listened to in Mapuche homes and communities touched on a variety of themes, from sentimental topics such as love and disappointment, male jealousy and death to aspects of daily life in the countryside and, of course, from time to time, more socially and politically engaged corridos and ranchera songs.

By addressing a wide range of themes "los Mexicanos" retained a vigorous presence on the airwaves after the right-wing military coup of 1973 and during the subsequent dictatorship and repression of Pinochet and his neoliberal capitalist cohorts. Prior to 1973, some politically Left-leaning radio stations had been pioneers in programming Mexican popular music and had a wide Mapuche and peasant audience. Pinochet's regime closed Radio Eleuterio Ramírez in Osorno and Radio Camilo Henríquez in Valdivia not because of their Mexican popular music programming but rather as reprisal for their affiliations with the Allende government.[13] Meulen's "Cancionero mexicano" continued on

Radio La Voz de la Costa, unaffected by political repression. Farther north, in the Province of Valdivia, "los Mexicanos" of the emerging Radio Austral gained a significant Williche and Lafkenche presence in rural and coastal areas respectively, as well as a widening audience in the working-class neighborhoods of urban centers.

On a more referential level, it is important to consider affinities between the ranchos of Mexico and the rural environments of Ngulu Mapu. Territorial resemblances are clear in the lyrics of "Juan Charrasqueado." By the second verse, listeners are situated at the contiguity of the rancho and the hacienda in Mexican geography. The hacienda represents the economic elites' large landholdings, being possessions mostly of a racially privileged White Creole sector of Mexican society. Ranchos in contrast range from small to midsize plots of Indigenous and mestizo farmers.[14] In "Juan Charrasqueado," the hacienda shines with a splendor—"the flower hacienda"—that contrasts with the rancho's "sad story" of a "ranchero in love."

The sentimental and socioeconomic geography of Mexican songs has its counterpart in the ancestral territories of the Mapuche People. The Chilean capitalist equivalent to the hacienda was the fundo, as it became known in what is today southern Chile (Bengoa 1990). As an economic, social, and racialized articulation of private property, the fundo expanded in parallel with the growth of a colonial and capitalist settler state and wealthy-class regime in Ngulu Mapu from the mid-nineteenth century. Historians Simon Collier and William Sater point out that in the mid-nineteenth century, hacienda ownership, until then dominant in the Central Valley of Chile, constituted "the clearest emblem membership of the national elite" (1996, 80–81).[15] After the defeat of the Mapuche uprising of the 1881 and as a result of more than two decades of intensive Chilean military deployment in what are now the provinces of Arauco, Malleco, and Cautín, the Chilean state's promotion of European immigration from the 1850s multiplied large landholdings in Mapuche territory. In 1860, three thousand Germans became part of this landowning economy (Collier and Sater 1996, 95).

After it crushed the Mapuche uprising of 1881, the Chilean state "decreed the Araucanía region as fiscal property." As it "auctioned off the lands, forming the fundos and the haciendas, developing the cities, a White society was organized in the region" (329–30). South to the Araucanía, the colonization of Ngulu Mapu included the Chilean state's geopolitical settlement in Williche territories of German settlers who were sponsored by the Chilean governments. The goal was to enable them to take "productive" control of native lands and extend the system of fundos. Between the provinces of Valdivia and Osorno, the fundos were mainly owned by German settlers, with some by Chilean Creoles and other non-Indigenous colonizers. Occupation of native land entailed the forced relocation of Mapuche families and communities into reducciones, a maneuver that was "legally" validated through new land title figures, such as the *Títulos de Merced* and other colonial forms of territorial subjugation.[16] As Chilean historian Luis Vitale describes, this post-1880 reality forced many Mapuches into agrarian servitude, creating a wage-earning workforce that had the choice to subsist as impoverished peasants on small or midsize parcels or to migrate to urban centers (2011, 430–31). Mapuche listeners, then, could readily identify with the Mexican socioeconomic geography of the rancheras and corridos. In Ngulu Mapu, the equivalent to the Mexican hacienda was the fundo; and the reducciones and/or parceled native lands mirrored the subalternized status of ranchos in Mexico.[17]

Within this social and territorial geography, the locus of the cantina, frequently referred to in rancheras and corridos, also has its counterparts in the environments of ranchos in Mexico and rural life in Ngulu Mapu respectively. With an accordion, a guitar, and improvised cries, Mexican popular music acquires aural preeminence in the cantina. Cantinas were part of the informal economy during the second half of the twentieth century and as such they played a formative role in rural areas. Rural cantinas were most often based in family homes that trafficked in locally produced chicha and alcohol from the city.[18] The association of cantinas, drunkenness, and music present in rancheras and corridos thus was inescapable in the countryside, especially among men. The cantina also played a vital role in the peripheral urban neighborhoods to which Mapuche migrants relocated. In the Rahue Bajo in Osorno, popular *chicherías* emerged as places that reconnected many Mapuches in the city to the environments of their birthplace or those of their kinship origin in rural Ngulu Mapu.

Mapuche poets of the late twentieth century Jaime Huenún and Juan Paulo Huirimilla emphasize the entanglement of the Mapuche taste for Mexican popular music with drunkenness.[19] In the last stanza of his poem "Cisnes de Rauquemó" (The Swans of Rauquemó) from his book

Ceremonias (1999), Mapuche-Williche poet Huenún portrays the journey of two Mapuche men through the rural environment of Füta Willi Mapu:

> Nos marchamos borrachos, emplumados de muerte,
> cantando unas rancheras y orinando en el viento.
> En mitad de la pampa nos quedamos dormidos
> cubriéndonos de escarcha, de hierba y maleficios. (Huenún 1999, 95)
>
> We went off drunk and feathered with death
> singing rancheras and pissing into the wind.
> We lay down to sleep in the middle of the prairie
> covering ourselves with frost, with grass and curses.

The end of Huenún's poem, with the drunken men who sing "rancheras and piss into the wind," leaves us with a scene of downfall and decay. The ranchera music is used here in the service of describing drunkenness and degradation and native land that has turned into a place of *maleficios* (curses). Juan Paulo Huirimilla, another Mapuche-Williche poet, also refers to Mexican popular music's resonance in the aural life of the Füta Willi Mapu. In his book *Palimpsesto* (2005), several poems are marked by music that tenuously sounds in the background as they make fleeting references to rancheras or corridos in settings linked to alcohol consumption such as the cantina.[20]

Not only on the radio and in poems but also for its locus as a site of relaxation and celebration, the buoyant atmosphere of cantinas in rural and urban Ngulu Mapu provided Mexican music with a large acoustic presence. Nevertheless, this dimension of Mexican popular music is just one reason for its aural traction among the Mapuche. The insistence on reducing the listening of this music to the setting of cantinas or chicherías has become a stereotypical fixation of certain discourses, especially in literature written by male Mapuche authors. However, focusing on this dimension nurtures a limited vision of the scope of Mexican popular music. It reinforces the tropicalization of the Mapuche subject within the colonial ideologeme of the "drunken Indian." Likewise, it endorses the Catholic priest's argument that such music spoke only of drunkenness and death.

In fact, the aurality of the rancheras and corridos has many and varied dimensions in Mapuche everyday life, as illustrated by the popularity of listening to "los Mexicanos" during breaks in the workplace or periods of free time at home. With the increasing availability of radio in rural areas of Ngulu Mapu, many Mapuche families and individuals, like my grandmother and uncles, enlivened their day by tuning in after lunch in the early or midafternoon. In the 1970s and 1980s, los Mexicanos" constituted the background music in the *micros* (minibuses) that transported people back and forth between urban epicenters and rural sectors. "Los Mexicanos" also played an important social and emotional role communicating between Mapuche living in the countryside and those in the city. Programs like Juan Meulen's "Cancionero mexicano" offered space for messages that ranged from announcements of events or community meetings to individual birthday greetings or messages from people in the city announcing a visit to family or friends in a rural area. Merged with the chords and voices of corridos and rancheras in the ether of Ngulu Mapu, the greetings and messages became part of the affective and communicative reverberation on the Mapuche ear.

Over the years, many conversations have led me to believe sentimentality is key to the bond between Mexican popular music and the Mapuche ear. Juan Meulen attributes the popularity among the Mapuche to a sonic aspect instead. To Meulen, "it's because of the rhythm; the rhythm that Mexican music has . . . because the people, I have realized, people from rural areas, do not listen that much to the message of the song but rather they like the rhythm" (interview by author, 2015). He recalls that radio listeners often requested tragic songs to accompany greetings on birthdays or saint's days or just as commonly would ask for rancheras like "Camino sin retorno" (A Road Without Return) by the Chilean group Los Manantiales that have sad lyrics to accompany joyous greetings or messages to relatives, fellow community members, friends, or loved ones. In his opinion, this oxymoronic practice of requesting sad or tragic rancheras or corridos for greetings of celebration or affection occurred simply because "people liked the rhythm." To Meulen, then, more than sentimentality and contextual associations between Mexico and Ngulu Mapu, what most captures and seduces the Mapuche ear seems to be the sonority of rancheras and corridos—the aesthetics at play, the way they sound and resonate, the fact that this is a music with simple rhythm, accessible to the aural and vocal senses of the Mapuche.

More recently, Mexican popular music has taken new shape in Mapuche aural life, especially among the youth. It is no longer music that only is heard or interpreted but now emerging Mapuche musicians have started composing their own rancheras and corridos, transforming the genre with Mapuche markers in content and form. Beginning in the early 2000s, the music group Werkenes del Amor (Messengers of Love) hails from the Lof Temulemu, a Mapuche community located in the Commune of Traiguen, Province of Malleco, at the heart of Ngulu Mapu. Los Werkenes del Amor have done outstanding work in composing corridos and rancheras that advocate for the Mapuche land-back movement, denounce settler colonial state repression and address other aspects of Mapuche life and culture. In re-creating/reinventing the older, foreign/Mexican forms through a Mapuche aesthetics and politics, the Werkenes del Amor have been certainly influenced by the tradition of "corridos revolucionarios" as well as by socially engaged singers of Mexican popular music, like Amparo Ochoa in the 1960s and 1970s and, more recently, the music of Los Tigres del Norte. A new life is beginning for los Mexicanos in Wallmapu!

RAP + MAPUCHE STYLE

We have seen how Mapuche youth in the 1950s and 1960s enthusiastically embraced the sounds of rancheras and corridos. By the 1990s and early 2000s, another generation, this one situated largely in an urban context, forged a new and very different auditory and musical bond with rap. Where it had taken many decades for Mapuche singers to compose their own rancheras or corridos,[21] rap took off as soon as it got to the ears of the Mapuche youth. In a decade, Mapuche soloists and bands were performing Mapuche rap.

Rap had first emerged in the mid-1970s in urban epicenters in the United States, such as New York and Los Angeles, among young people from Black communities faced with racism and class segregation (Fernando 1999; Higa 1999; Morgan 2009; Perry 2004; Rose 1994). Likewise, when it reached Chile and Ngulu Mapu in the mid-1980s during the Pinochet dictatorship, rap and hip-hop culture initially gained strength in cities, which were fertile ground for the restlessness of youth chafing against the authoritarian regime's censorship and its restrictions of access to public space. In fact, massive street protests against the dictatorship's human rights abuses and neoliberal policies had erupted throughout Chile in 1983. As young people from socially marginalized and impoverished social sectors of greater Santiago began to channel their rebellion through rap and hip-hop, they often employed the aesthetics of this new form of expression to give voice to their protests.

Guillermo Navarro Cofré, one of the hip-hop promoters, notes that, the "interventions of this incipient movement did not have political content," but more generally were comprised of dissatisfied youth aiming "to stop being invisible" (2023, 25). However, I would add to this that although their lyrics were not directly "political," the mere practice of gathering in urban plazas, parks, street corners, or alleys of the city, as hip-hop and rap followers did, constituted a rebellious act by defying the authoritarian regime's restrictions on "public activity" in the country.[22] Whether these gatherings occurred in downtown areas of cities or in "la pobla," that is, in the neighborhoods of peripheral and working-class barrios (Meneses 2014, 10–21), the active use of urban public spaces by hip-hop musicians and their followers, without asking for approval from governmental authorities, had a clear political edge.

For this new generations of young people in Chile, like their counterparts in North America, hip-hop rapidly became a way of inhabiting urban space and forming their own "acoustic territories" (LaBelle 2010), or modes of acoustic habitation in cities. A sense of territory was at the heart of narratives in this movement as young hip-hoppers set up circles, installed their cassette players, practiced breakdancing, and competed among themselves in urban spaces, such as plazas, alleys, or street corners in working-class and peripheral barrios as well as at other downtown spots in the cities. Lalo Meneses's *Reyes de la jungla: Historia visual de Panteras Negras* (2014) documents how they took over downtown Santiago's Bombero Ossa passage (known as "the alley") made it into an emblematic meeting point. Many of the young people who joined this rhizomatic movement were from urban areas impacted by the increasing precarity of social and material living conditions that vast segments of the population experienced under the market economy–oriented monetarist policies imposed first by Pinochet and then embodied into the framework of a neoliberal democracy.

In Santiago, many Mapuche families resided in the peripheral urban areas, such as the communes of Pudahuel, Renca, Peñalolén, and La Florida, where hip-hop circles took shape in the 1980s. At the same time, throughout Ngulu Mapu, in cities such as Concepción, Temuco, or Valdivia, hip-hop and rap also were gaining significant aural appeal among young people, including the new generation of Mapuche. According to Santiago-based Mapuche rapper Waikil (Jaime Cuyanao), this was the case, for example, in Población Lanin, a barrio located in the populous Pedro de Valdivia sector of the city of Temuco, where a Mapuche hip-hop group, the Brocas de las Naquis (Kids on the Corner), had already formed in the late 1980s.

For the young Mapuche people who ventured into rap, the work of earlier hip-hop artists in Chile undoubtedly was influential. The Brocas de las Naquis, which constituted a milestone as a rap group emerging from the heart of the Araucania region, established significant ties to the Mapuche political movement in Temuco.[23] Meanwhile, back in Santiago, by the early 1990s, the iconic hip-hop group Panteras Negras had also become a notable force associated with the country's countercultural vocal, musical, and political scene. Voices from the African American North, heard via VHS videos, cassettes, and DVDs, were equally influential. For young Mapuche, rap became a potent vehicle for expressing dissatisfaction, a parallel to its impact among Native American youth in what is now called the United States where, in the words of Seminole historian Kyle Mays, "Indigenous hip hop" became, "through cultural expression, an assertion of Indigenous sovereignty" (2018, 21).[24] In short, rap gave a voice to people who had not had one.

Within the Mapuche diaspora in greater Santiago, one of the groups that gained a voice and started Mapuche rap was Wechekeche Ñi Trawün (Mapuche Youth Gathering). It emerged in the Commune of La Florida around 2000, defining itself not so much as "an artistic group" as "a grassroots organization" (Millaleo, interview by author, 2024). Wechekeche Ñi Trawün was comprised of young Mapuche men and women who had grown up in Santiago. Its founding members were Paul Paillafilu (Filutraru), Ana Millaleo (Millako Leufü), Jaime Cuyanao (Waikil), Axel Paillafilu (Kallfulafken), Ximena Painemal (Kallfuray), Daniel Millapan, and Cristian Cuyanao.[25]

As members of the Mapuche diaspora, a key goal for the group Wechekeche ñi Trawün was to generate connections among the young Mapuche in greater Santiago to reconstruct their sense of peoplehood belonging and of being part of the broader context of Wallmapu. This led them to organize group trips to southern Chile to reconnect to the native lands in Ngulu Mapu their families came from, a practice that they continue today. In January and February 2004, on a trip to Temuco, in the urbanized and colonized heart of the Ngulu Mapu, as they had

FIGURE 5.2 · Wechekeche ñi Trawün. Photo taken in the Jardín Intercultural Antü Mahuida, Commune of La Florida, Santiago, Chile, 2018. Leading members of the group: (*left to right*) Kalfullüfken Paillafilu, Valentina Kurin, Filutraru Paillafilu (with trutruka), and Ana Millaleo. Courtesy of Paul Paillafilu.

to figure out how to make money to finance their journey, they turned to music as a means of fundraising while also presenting themselves before the community. They composed a piece that mixed vocal and sound elements of rap with others from the ül tradition and Mapuche musical instruments. Even though they ended up not performing what they composed, this exploration reflected well an important dimension of what they wanted to put in practice. In this regard, as group member Ana Millaleo remarks, "since our inception, we consider ourselves as a group of Mapuche fusion music" (interview by author, 2024).

The mixture of musical registers conforms to Wechekeche ñi Trawün's desire to give an account of Mapuche listening, both in the countryside and in the city, where various musical genres coexist and at times mingle in Mapuche aurality. As a consequence of this aspiration, the albums they went on to produce in the first decade of the twenty-first century include a rich mix of hip-hop, salsa, cumbia, rock, soul, and ranchera, among others. By the late twentieth and early twenty-first centuries, as I have previously pointed out, rancheras, corridos, romantic ballads and cumbia songs were highly popular in rural areas of Ngulu Mapu, as well as, especially in some Mapuche-Williche areas, Chilean cueca also had a significant popularity. Hence the grounding aesthetics of this new Mapuche fusion music is reflected in how it is always imprinted with a touch of what, in one of their songs, they refer to as "Mapuche style."[26] The albums Wechekeche Ñi Trawün released in their first decade, such as *Wechekeche Ülkantun* (2005), *Wechekeche Ka Kiñe* (2006), *Kuifikeche Ñi Trawün* (2006), *Wallmapuche / Wajmapuce: Ka Puel Kona* (2009), and *Que viva la Raza* (2009), were released and circulated on CD. Later, around 2010, they jumped to digital platforms such as YouTube and Spotify.[27]

Rap music and hip-hop culture loomed large in Wechekeche Ñi Trawün's juxtaposition of genres from the time of the collective's emergence in Santiago in 2000. Ana Millaleo explains, "We started making rap, because many of us came from hip-hop groups" (interview by author, 2024). They borrowed sampling, scratching, beat boxing, rhyme, repetition and flow, and sometimes improvisation, all of which they used in producing rap songs with a Mapuche linguistic, cultural, and political inflection. Overall, then, their "Mapuche style" was always more than an aesthetic expression: It was part and parcel of their intentional use of music as a medium for engaging the Mapuche youth in the struggle of Wallmapu as a whole. As Ana Millaleo put it, "In a time in which [Chilean] social media did not exist yet," music was envisioned as "a tool of communication; firstly it was for conveying what we felt, what we witnessed, what happened to us, and then to communicate what people wanted us to say and that this be known in the whole territory" (interview by author, 2024).

Wechekeche ñi Trawün helped spur the development of what could be called *Mapuche rap*—music originating in urban environments but strongly situated in the linguistic, cultural, territorial, and historical struggles of the Mapuche people.[28] Mapuche markers include the trope of the Mapu, insertion of Mapudungun, vocal and sonic traces of the long native tradition of ül (chant), and the use of Mapuche musical instruments, such as the trutruka, the kull kull, the kultrun, the pifilka, the *kaskawilla*, and the trompe.[29] In Mapuche rap, elaborate lyrics coexist with the improvisation and spoken word characteristic of rap. Linguistically, Wechekeche Ñi Trawün's rap songs often alternate use of Mapudungun and Spanish in their aim to appeal both to Mapuche who know the Indigenous language and the vast community of Mapuche who are not speakers of Mapudungun.

A good example of their rap music is the song "Mapudungufinge" (Speak Mapudungun!), which they composed around 2014 (Wechekeche Ñi Trawün 2015). In "Mapudungufinge" they used the native language with the objective to support the Mapuche demand for the recognition of Mapudungun as an official language in the Province of Cautín. The song starts with the sounds of scratching, followed by juxtapositions of scratching with sampling from an ül (Mapuche chant) interpreted by well-known ülkantuchefe Sofía Painaqueo,[30] after which the voices of Wechekeche Ñi Trawün go on,

> Mapudungufinge!
> Ta iñ nütram alkutunge
> Kom pikeyin, yewekilnge!
> Trepepe ta mi piuke. [*repeated*]
> Kuyfi mew ta iñ kewüṅ pewütuy ta iñ Mapu mew
> Pu kuyfike che em allkütumekefuy engün
> itro fill mongen, üñüm ñi ülkantuael, kullin ñi dunguael.
> Femngechi pelontukefuy engün
> mülerkefuy kiñe fütra dungun, kay kiñe fütra kimün,
> fey azümürkefuy engün chumngelu Mapuzüngun pingelu,
> chem mew am Mapuche pingeyiñ.
>
> Speak Mapudungun!
> Listen to our conversation
> We all say don't feel embarrassed!
> Wake up your heart!
> In the old times our language sprouted from our land
> our ancestors listened to it carefully,
> to all forms of life, the songs of the birds, the voice of the animals.
> So, they had the clarity
> to know a great language existed, and great wisdom,
> They understood that Mapudungun was its name,
> and why we are called Mapuche.

Instrumentally, "Mapudungufinge" includes Mapuche wind instruments like trutruka, kull kull, pifilka, and kaskawilla, and the percussion instrument, kultrun, along with non-Indigenous sound technologies such as synthesizer, keyboard, electric bass, and sounds electronically generated and mixed in the studio. This is a choral rap. What gives a common thread and collective vocal flow to the whole is the chorus, "Mapudungufinge! / Ta iñ nütram alkutunge / Kom pikeyin, yewekilnge! / Trepepe ta mi piuke!," which is sung by the entire group and repeated four times. Multiple male and female Mapuche voices alternate in rap, recitation, and song.[31] This is evidenced with greatest clarity in the staging for media platforms such as YouTube. Faces and bodies of Mapuche young people mingle with those of elders and children, exemplifying cross-generational community relations. The Wechekeche Ñi Trawün members dress in a mix of Western-style clothes adorned with characteristically colorful Mapuche wardrobe and ornaments, such as trarilonko (headband), makuñ (poncho), and, in the case of women, *trapelakucha* (made-of-silver breast ornament). Visually, the green, red, yellow, black, and white colors of the Mapuche flag that has widely identified the Mapuche movement since the early 1990s tend to prevail throughout the video. The combination imbues the poetically well-crafted lyrics of "Mapudungufinge" with a powerful resonance. In this way, their performance illustrates African American scholar Imani Perry's notion that rap constitutes "a mixed medium" that "as an art form" always combines different elements (2004, 38)[32]—in this case, poetry, vocality, sound, visuality, bodies, and territorialities, all of which overlap through the choral performance of the song.

Wechekeche Ñi Trawün's creative amalgamation also depicts the multiple territorial, cultural, and historical relationships that being part of a Mapuche diaspora entails in a colonial urban environment. This what they aimed to convey in the visual accompaniment for the song "Mapudungufinge" when they issued it on YouTube in 2015. Throughout the video, we are treated to a moving collage of disparate settings in greater Santiago, ranging from a plaza in the Villa O'Higgins neighborhood in the Commune of La Florida to the corporate glass skyscrapers of the financial district, to the Bosque Panul, a forest in the pre-cordillera area, where they are accompanied by a Machi. Visually and sonically, this illustrates the historical experience of Mapuche in a wingka city, complicating the urban/rural divide and positioning creative ways to spiritually and politically reconnect to the Mapu.

In this sense, "Mapudufinge" is about the native language as a connector to a greater sense of Mapuche Ngen (being Mapuche) by highlighting links between the practice of Mapudungun, the Mapuche Kimün (the Mapuche knowledge system), and the Itro Fill Mongen (the diverse ecosystem of lives of the Mapu). As the song points out in the penultimate stanza of "Mapudufinge" and before repeating the initial stanza, the Mapuche language is the expression of the Mapu as a whole,

> Mawida, wanglen, kürüf, kütral dungukey,
> Tralkan, llefken, ƚeufü, ƚafken, inchiiñ emu nütramkülei.
> Mapuche dungun mew piuke mew puwtukey,
> Mapuche dungun mew ta iñ ülkantun tripaley.
>
> The mountain, the stars, the wind, the fire speak.
> The thunder, the lightning, the river, the sea, converse with us.
> Also the people of the coast, the rivers, the sea, the country speak.
> Speaking in the Mapuche language always touches the heart,
> We sing our songs in Mapudungun.

In both their cultivation of rap and what they call "Mapuche fusion music," Wechekeche Ñi Trawün intertwine multiple dimensions of Mapuche Ngen, even though they are mediated by the colonizing entanglements of the Chilean wingka city; in this case, Santiago. Sense of community, revitalization of the language, desire to reconnect with the Mapu and the territorial struggles in the Ngulu Mapu—and more widely in the Wallmapu—infuse their varied musical creations. As Perry points out on black urban life in the United States, rap is expressive of "not only aesthetics" but also, above all, "a politically charged identity" (2004, 26–27). This political sensibility is what has also been central to the sonic life of Mapuche rap, which for many young Mapuche constitutes a critical tool to generate their own modes of acoustic habitation in wingka cities.

ÜLKANTUN, AN ENDURING SOUND

Ül is synonymous with song. The verbs *ülkantun* or *ülkatu* refer to the action of singing. According to Mapuche pedagogue and researcher Ramón Curivil Paillavil, ülkantun constitutes a medium through which "Mapucekimvn and Mapucerakizuam are transmitted" (2002, 22).

The song expresses aspects of kimvn or kimün, the Mapuche system of knowledge, as well as rakizuam or rakiduam, terms that denote the thought of the Mapuche people. Consequently, ül and its practice, ülkantun, are linked to various dimensions of Mapuche life, to quotidian activities such as planting and harvesting, as well as to higher religious and spiritual matters and to love. As a distinct musical genre, ül takes shape through different subgenres, characterized by important stylistic, thematic, and contextual variations. Curivil Paillavil defines ten types of ül.[33] Elisa Avendaño Curaqueo, whose ül practice will be discussed by the end of this chapter, warns, "People who are ignorant of Mapuche music generally believe that the genre is one and homogeneous and that is not the case. Mapuche music is as fun as it can be very monotonous because it depends a lot on where we are, what social moment we are in, or what political moment we are in" (interview by author, 2023). As Avendaño Curaqueo suggests, ül entails an overarching, flexible approach to music-making used to orally and vocally express the various tangible and nontangible experiences and life situations of people, communities, and territories in relationship with the Mapu.

Ül has numerous functions in Mapuche life. It is a musical language that facilitates communication between people and communities—a Mapuche conversation can involve singing. It can enliven and brighten daily life. In certain critical circumstances, ül helps deal with trauma and pain and, in that dimension, it assumes a medicinal role. It gives us *newen* (spiritual and physical strength) during war and struggle. On a religious and cultural level, it helps cultivate spiritual and symbolic relationships with the tangible and nontangible environment of the Mapu. There are ülkatu or ülkantun that constitute the exclusive domain of whoever is the Machi, our Mapuche medicinal and spiritual authority. Finally, employing voice and sound, ül expands and enriches the expressive capacity of Mapudungun, not only communicatively but also aesthetically. As art form, then, ül cultivates the sense of beauty and pleasure in Mapuche culture.

The land dispossession and forced relocation that many Mapuche suffered in the late nineteenth and early twentieth centuries left its mark on the vocal culture of Ngulu Mapu. Chants that expressed the predominant mood of sadness and melancholy proliferated across different territories. When Catholic missionaries and secular researchers, mostly of European origin, traveled into native communities to collect the ül, this is what they found and what they came to consider typical reflections of Mapuche life. An important figure in this endeavor was the Capuchin priest Félix José de Augusta, a missionary and surgeon of Jewish German origin, who mixed his activities of Catholic evangelization with the study of the language.[34] Around 1909 and 1910, de Augusta traveled to Ngulu Mapu, visiting the reducción of Wapi, in the Lake Budi area; and then, southward, he went to Panguipulli. He gathered tales, letters, chants, and narratives of *costumbres* that he published under the title *Lecturas araucanas* (1934). De Augusta visited the Mission of the Panguipulli twice. On his last visit, in 1910, he traveled with a phonograph, even though most of the ül were put as written transcriptions and he recorded only "some songs and an excerpt from a parliament" (de Augusta 1934, iii).[35] In *Lecturas araucanas*, he reproduces his transcriptions in Mapudungun, along with translations of each song into Spanish.

Listening to the colonial archive is always a critical challenge. Indeed, going through the many dozens of songs de Augusta collected, one can note the biases of his colonial ear through the way he arranged his selection to emphasize songs related to alcohol consumption, love stories or entertainment among the Mapuche. This folkloristic overtone of his ül selection leaves the impression that Mapuche music ignored the painful historical situation in which the Ngulu Mapu was immersed. A more likely explanation could be that, for self-protection, the Mapuche chose not to sing more historically and politically charged songs before a settler Christian missionary.

However, one ül in de Augusta's selection from his visit to the Mission of Panguipulli stands out as a song that tells us a lot about the historical moment in which the Mapuche were immersed. Under the title in Spanish "Se conforma" (A Resigned Man), the Capuchin priest transcribes it into his own Mapudungun alphabet, accompanied by his version in Spanish. Here I copy de Augusta's transcription and translation as he published them in 1934. Then, since I did not find completely satisfactory de Augusta's translation, I offer here an alternative-version Spanish—drafted in collaboration with Claudia Ingles Hueche and Pablo Millalen Lepin—besides my translation into English,

Kuifi mai küme wentru ŋefun em;
L'ai ñi pu loŋko-em, pofre, ŋewen.
Mëleyei ŋa Mapu tañi cheŋemom:

Anüñma-anüñmanieyen ñi Mapu.
Ñi Palŋiñ-Mapu ñi cheŋemum:
Femŋen mai ñi pofre ŋewen meu,
Welu chumlam.
Ayüeneu ñi inátuateu pu wiŋka,
Inche chem kulpa nielan.
¿Chumafun?
Ngënechen mai kimi ñi chumateu.

Antes, pues, era yo hombre bien puesto;
Por la muerte de mis caciques he quedado pobre.
Tengo terrenos y había sido rico en ellos:
Vivo de estable aquí,
En mi tierra de Palniñ, donde yo era rico.
Cierto que me he quedado pobre,
Pero no he hecho mal.
Me quieren echar de mi tierra los huincas.
Pero yo no tengo culpa en eso.
¿Qué se habrá de hacer?
Dios sabe lo que dispondrá de mí.
(de Augusta 1934, 158)

Antiguamente, pues, era un hombre pleno.
Cuando murieron mis *pu longko*, quedé pobre en mi ser.
Había tierra para poder ser persona con espiritualidad.
En mi tierra estoy bien "aposentado" siempre.
En mi tierra de Palniñ, yo era persona en plenitud.
Por eso, pobre de espíritu ahora,
pero no he hecho nada mal.
Los *wingka* me hostigan por mi tierra.
Pero yo no tengo ninguna culpa.
¿Qué voy a hacer?
Ngënechen sabe nomás qué hará conmigo.

In the old days, well, I was a fulfilled man.
When my *pu longko* died, I became poor in my being.
There used to be [enough] land to be a spiritual person.
On my native land, I'm always well "put up."

On my land of Palniñ, I was a whole person.
Though it's true I'm now poor of spirit,
I haven't done anything wrong.
The wingka people harass me for my land.
But that's not my fault.
What can I do?
Only Ngënechen knows what he will do with me.[36]

According to de Augusta, this emotionally charged ül was sung by Carmen Kumillanka Naqill, from the rural community of Palniñ, probably close to the Panguipulli Mission. As we read and "hear" from the lyrics, the woman ülkantuchefe impersonates a Mapuche man who feels "poor" in spirit, or as an impoverished being, due to the passing away of his pu longko (literally, "the community heads"). This *wentru* (man) also tells us about the bothering presence of settlers (*pu wingka*) around his native land, located in Palniñ. The way I listen to this ül is a testimony of a time in which the Mapuche self-governance system is being dismantled in many areas of Ngulu Mapu due to the ongoing Chilean colonizing process, including Catholic missionization of communities. Interestingly, de Augusta comments in a footnote that the speaking persona of this song may have become "poor" because of "theft of his animals," "a bad harvest season," or "conflicts" that unfavored him (1910, 158). Nevertheless, the lyrics in Mapudungun does not use exactly *rico* and *pobre* in a restricted economic sense; rather, the persona of the song speaks more of how his very subjectivity is affected and impoverished in a deeper sense, in his *ngen* (Mapuche being; personhood) and *newen* (spiritual and physical energy). In a relational way, the land has meant and continues meaning a lot for the persona of this ül. Of course, a Christian missionary fails to acknowledge that the Mapuche have their own spiritual, religious, and ontological views that are interwoven with land relations. Contrary to the reductionist understanding of this ül that de Augusta offers, it seems clear to me—from a deeper Mapuche perspective—that this song beautifully conveys that circumstances of colonization not only affect Mapuche lifeways in socioeconomic terms but also undermine land relations and being Mapuche in a spiritual, ontological, and epistemological dimension. The "impoverishment" of the Mapuche that Chilean colonialism brought with it into Ngulu Mapu was of multilayered relational impact, namely, economic, social, political, spiritual, ontological, and epistemological. As a song that emerges from

FIGURE 5.3 · Elisa Avendaño Curaqueo and her kultrun. Photo taken during a recording session for one of her ül albums, *Raíz Mapuche* (Mapuche Roots). Curacautin, Ngulu Mapu, 2005. Courtesy of Elisa Avendaño Curaqueo.

a specific personal and collective life situation, is a beautiful and profound statement about what good land relations and its counterpart, the invasiveness of colonial agents, has meant in the history of Ngulu Mapu.

With its powerful content, this early twentieth-century ül also illustrates key characteristics of the genre. First, an ül tells a story involving a life situation affecting a Mapuche person or community. In this case, as in much of the present-day practice of ülkantun, the song denounces a situation affecting the Mapuche being in a context of colonization and crisis of Mapuche society. Second, it also communicates strong emotions; it moves us. Traditionally, ül exist as a vehicle for expressing feelings linked to the Mapuche or the Mapu. As Curivil Paillavil puts it, all ül songs "seek to reach the heart" (2002, 23); and in Kumillanka Naqill's song, the sadness and melancholy of a "resigned man" inevitably touches us.

The cultivation of ül has continued among the Mapuche throughout time. This brings us back to one of today's chief practitioners, Elisa Avendaño Curaqueo,[37] who, since the 1960s has been active in the practice of Mapuche music as well as always engaged with the struggle of Mapuche communities to recover lands that settlers had taken away from them. Elisa Avendaño Curaqueo began her practice of Mapuche chanting around 1966, when she was just ten and a Machi happened to offer her a chance for the first time to play the kultrun—the main percussion instrument in Mapuche life. Ever since then, she has relied on the kultrun in her cultivation of ülkantun.

With the nasal voicing, which is a traditional characteristic of ül's vocal style, Avendaño Curaqueo has revived and created several ül. One of them, "Celinda" (also written as "Selinda") has been part of the soundscape of Ngulu Mapu for decades, existing in multiple versions and interpretations. It tells the story of Celinda who, soon to be married, happily prepares to leave her lof (local territory) for the urban settlement of Chol Chol, to buy new clothes. In Avendaño Curaqueo's interpretation, the ül begins,

Celinda anay, Celinda
Celinda anay, Celinda.
Müna ayüi, Celinda.
Müna ayüi, Celinda.
Füta ngealu, Celinda.
Füta ngealu, Celinda.
Müna ayüi, Celinda.
Müna ayüi, Celinda
Celinda anay, Celinda.
Amuyea-alu nga
Chol Chol waria, Celinda
Amuyea-alu nga
Chol Chol waria, Celinda
Füta ngealu, Celinda
Füta ngealu, Celinda
Müna ayüi, Celinda
Müna ayüi, Celinda
Chol Chol waria amualu
Celinda anay, Celinda.
Chol Chol waria amualu
Celinda anay, Celinda
Celinda anay, Celinda
Celinda anay, Celinda
Müna ayüi, Celinda
Müna ayüi, Celinda

Füta ngealu, Celinda
Füta ngealu, Celinda
Chol Chol waria, Celinda
Chol Chol waria, Celinda
Ka ngillelalngealu
trarilonko, Celinda
Ka ngillelalngealu.
trarilonko, Celinda.
Müna ayiui, Celinda.
Müna ayiui, Celinda
Ngillelalngealu.
trarilonko, Celinda
Ka ngillelalngealu
kiñe chamal, Celinda
Ka ngillelalngealu
Kiñe chamal, Celinda
Celinda, Celinda,
Celinda, Celinda

She is very happy, Celinda
She is very happy, Celinda
She will get married, Celinda
She will get married, Celinda
She is very happy, Celinda
She is very happy, Celinda
Celinda, Celinda
She is going
To the city of Chol Chol, Celinda
She is going
To the city of Chol Chol
She will get married, Celinda
She will get married, Celinda
She is very happy, Celinda
She is very happy, Celinda
To the city of Chol Chol, she will go
Celinda, Celinda
To the city of Chol Chol, she will go
Celinda, Celinda
Celinda, Celinda

Celinda, Celinda
She is very happy, Celinda
She is very happy, Celinda
She will get married, Celinda
She will get married, Celinda
In the city of Chol Chol. Celinda
In the city of Chol Chol, Celinda
Also her kin will buy
a headband for her, Celinda
Also her kin will buy
a headband for her, Celinda
She is very happy, Celinda
She is very happy, Celinda
She will buy
a headband, Celinda
Also her kin will buy
one chamal for her, Celinda
Also her kin will buy
one chamal for her, Celinda

This ül celebrates the emotional and social trajectory of a Mapuche woman as she moves toward a new stage of life. Historically, ül are performed with the accompaniment of traditional Mapuche instruments, and Avendaño Curaqueo often performs this one accompanied by the trompe, the equivalent of the Western jew's harp, a small instrument played while held in the mouth. The timbre of the trompe doubles the song's lively mood while listeners follow Celinda's happy story:

Füta ngealu, Celinda
Füta ngealu, Celinda
Müna ayüi, Celinda
Müna ayüi, Celinda.

Entertaining and amusing, the song represents an example of the *ayekan-ül*, a subgenre of ül that, as defined by Avendaño Cuarqueo, aims "to animate" its audience (interview by author, 2023). Avendaño Curaqueo's high-pitched voice allows the song to reverberate. Its intensity moves Mapuche and non-Mapuche audiences alike, whether they understand the

Mapudungun lyrics or not. The resource of repetition, a common feature of ül, augments the reverberations; but the vocal and expressive strength of the ülkantuchefe, which involves mouth, throat, and body, ends up affecting "the heart"—the blood and aural flow—of the listener.

The story takes a leap in the third stanza: "Amuyea-alu nga / Chol Chol waria, Celinda." In Chol Chol, they will buy her ornamental clothing, her trarilonko and her *chamal*. The reference to Chol Chol places us in a particular territory of the Ngulu Mapu: Nagche Mapu (territory of the people of the plains). Between iterations and variations of its verses, the song continues to develop through a sequence of stanzas and culminates by confirming the joy and emotional stimulation that Celinda finds:

Umawtuwealay, Celinda
Umawtuwealay, Celinda
Müna ayüi, Celinda
Müna ayüi, Celinda.

Before Elisa Avendaño Curaqueo's time, "Celinda" had mostly been performed by ülkantuchefe men. As a woman chanter, she endows a song of a Mapuche woman's self-realization with a particularly effective vocal force and sense of celebration. Avendaño Curaqueo has cultivated other variants in her ülkatu, honoring the principle that ül is a genre that tells of many types of situations in Mapuche life. Drawing on her active involvement in the territorial struggles of the Mapuche people since the 1960s, she has employed the genre to address political matters that touch the community. Indeed, she tells me, her first public presentation as ülkantuchefe was in Temuco in 1980 at one of the inaugural events in Temuco of Ad Mapu, one of the most important organizational references of the Mapuche movement at the end of the 1990s.

"Wilipang" (also written as "Wilipag") is another ül that expresses its commitment to our people. Around 1995, amid the "democratic transition" in Chile, Avendaño Curaqueo composed this ül to honor the memory of Wilipang (Uña de León, or Lion's Claw), a young Mapuche leader of the fight to recover Mapuche land who was murdered by the Chilean state police during the socialist government of Salvador Allende (1970–73). Because of her personal connection to Wilipang, her voice in this highly political song takes on a markedly emotional and personal resonance. Here, the first part of "Wilipang,"

Wilipang nai, Wilipang
Kiñe antü nga nütütuaiñ ta iñ Mapu nai
Pienu . . . eimi nai lamngen
Welu eimi nai, Wilipang
Lofnafemngey eimi tralka anai, Wilipang
Wilipang nai, Wilipang
Fei nga küleyetuy ta mi mollfüñ nai, lamngen
Umawnagkunungeyeimi nai Wilipang
Ngümagümangetuy ta mi pu che nai, lamngen
Che mew nga feletuy nga ta ñi lamngen
Ingaküleyetuy ta mi mollfüñ nai Wilipang
Wilipang nai, Wilipang
müna kutrankawuwkületui ta ñi piwke nai, lamngen.

Wilipang, Wilipang,
Wilipang, Wilipang.
"One day we will recover our land,"
you said to me . . . , brother.
But you, Wilipang,
they knocked you down with one shot, Wilipang.
Wilipang, Wilipang,
that is why your blood is dripping, brother.
They left you lifeless, Wilipang.
Your people are crying, brother.
Why are you like this, brother?
Your blood is spilled, Wilipang.
Wilipang, Wilipang,
My heart is in so much pain, my brother.

Although she sometimes accompanies it with the kültrun, a Mapuche drum, Avendaño Curaqueo composed this ül to be performed a cappella. It is an elegy, a lament, a tribute to the figure of a *weichafe* (Mapuche fighter), and a memorial. In its elegiac tone, the lyrics continues for about nine more verses, the last three ending with the repetition of Wilipang's name. With repetition and apostrophic allusion, Avendaño Curaqueo revives Wilipang and makes him present. She calls attention to the fact that there are many Wilipangs, young Mapuche leaders murdered by repressive agents of the Chilean state and, considering that "Wilipang" refers to Allende's socialist government, notes that "Mapuche have been murdered in every type of government in this Chile, not only in the Bachelet period, not only with Piñera, not only with Pinochet. We Mapuche have been murdered."[38] She adds, "Every time there is a Mapuche movement, every time the Mapuche move to recover their land or fight against injustice, that great leader has always been assassinated. It is the only way to achieve economic power, political power, the power of the Chilean state, to govern more calmly" (Avendaño Curaqueo, interview by author, 2023).

The cultivation of ül acquires a particular political force in the vocal and instrumental art of Elisa Avendaño Curaqueo. Part of this derives from the accompaniment with traditional Mapuche instruments, for example Avendaño Curaqueo's use of the kultrun, which provides an immediate connection to the Mapu, or, as she puts it "Kultrun is what gives me energy; and if my kultrun beats, my heart beats . . . it jumps like my heart, like my *piuke* . . . that is, the kultrun has to sound like my heart sounds, and at the same time the kultrun resonates with the earth" (interview by author, 2023). However, for all performers, whether with or without the kultrun, the ül is not only an art of the (human) voice but also an indelible part of the greater sound of the Mapu.

CONCLUSION: YESTERDAY, TODAY, TOMORROW

May 2024. It is a wintry Wednesday afternoon in Osorno. I spent the afternoon at the house of the kuifikeche Juan Meulen, in the town of La Esperanza. We have talked at length about the life of the rancheras and the corridos in the Füta Willi Mapu. I take a "collective taxi" back to where I'm staying in the center of Osorno. The driver has the radio on and is listening to a Los Mexicanos program. The format is the same as when Juan Meulen started his program in 1971, alternating songs and messages.

Two weeks later I am in Temuco. It's Monday, May 27. I move to the urban center of Padre de las Casas, bound for the rural sector in the area of the Paillanao Community, where, together with the collaborative team of which I have become a part, we will work all morning at the home of the ülkantuchefe Elisa Avendaño Curaqueo to develop a project for the recovery and revitalization of Tayül-Ul. Once this meeting is over, the ülkantuchefe offers to drive me to Chol Chol, approximately an hour and a half from Padre de Las Casas, where I will visit Clara Antinao Varas, a Mapudungun pedagogue who cofounded the Mapuche radio program Wixage Anai in Santiago in the early 1990s. On the way to Chol Chol, ülkantuchefe Elisa Avendaño Curaqueo plays a series of ül on the radio. She also sings some of those songs to me. All this has happened in less than two weeks. It is the contiguity and superposition of the music of Ngulu Mapu.

....

In the Mapuche worldview, time is formulated as a continuum of interwoven, overlapping cycles. Its cyclical nature invokes circulation, renewal, and even repetition. The Mapuche do not divide time into periods, nor do they distinguish between new and old. For instance, we do not celebrate a "new year." By the winter solstice in the southern hemisphere—late June in Wallmapu—we celebrate Wiñol Tripantü, that is, the renewal or rebirth of life in the Mapu. More literally, it refers to "the return of the sun."[39] In this cyclic conception of time, the old and the new intersect, repeating but also renewing. *Wüya* (yesterday), *antü* (today), and *mañana* (tomorrow) are cardinal references in the interconnected time of the Mapu. Mapudungun contains no specific words for "past," "present," and "future," nouns that, in Western conceptions of time, are directly tied to linearly progressive, abstracting macro-narratives of what is then called "History."

A Mapuche sense of temporality underlies the structure of the present chapter, in which Mexican popular music, rap, and ül overlap, conforming to the same arc of sonic contemporaneity in Ngulu Mapu and more extensively in Wallmapu. I have formulated this final chapter as an invitation to listen to rancheras and corridos, rap, and ül within a temporal circle, all of them being musical registers that sound contemporary to the Mapuche ear.[40] Acknowledging this flow of temporalities in Mapuche life necessarily leads us to what scholar Juliana Hu Pegues's elaborations on "space time" in Native Alaskan positions as a form to "refuse the colonial temporal binary of traditional versus modern" (2021, 154). Regarding Mapuche musical practices, in this chapter I have intentionally avoided a taxonomical schematization of "new" music versus "ancestral" chanting, or "modern singers" versus "traditional chanters." The aural life of the ül is as contemporary as rap or *Los Mexicanos* in the sound, vocal, and listening experience of Ngulu Mapu. This palimpsest of *musics* that overlap in Ngulu Mapu cannot be assimilated into the logic of a multiculturalist "audiotopia" (Kun 2005). Rather, it constitutes a heterogeneity of musics that battle amid the fraught history that Mapuche live as a colonized and racialized people, mediated by the predatory omnipresence of settler neoliberal capitalist forces on the lands of Wallmapu.

Mapuche creativity interferes with the dominant colonial script by keeping alive the continuum of vocal and instrumental arts of centuries, such as the ül; or linguistically and culturally resituating recent non-Mapuche forms of arts, such as rap. On the one hand, at stake in singing these musics among the Mapuche is what scholar Licia Fiol-Matta has formulated regarding the status of voice as "a producer of thought" (2017, 8). On the other hand, the experience of *los Mexicanos* in Ngulu Mapu is a telling example of what musicologist Ana María Ochoa Gautier characterizes as the generative ontological and epistemological force of "aurality" (2014, 34). Listening to rancheras or corridos, just as to ül or rap, for many Mapuche has been a mode of feeling, thinking, and acoustic habitation in a sonically and historically colonized environment.

Collective memory somehow works in the domain of the musical and sonic. On a preliminary hearing, what simply seems to be disparate musical and generational tastes among the Mapuche shows some striking patterns that resonate across the different genres analyzed in this chapter. For instance, within its different contexts, corridos and the ül share the common trait of being musical narratives that tell us a story. Scholars of Mexican popular music have pointed out that a corrido is a form of storytelling, constituting "an epic-lyric-narrative genre" that "tells about those events" that "powerfully" affect "the sensibility of the multitudes" (Mendoza [1954] 1976, ix). Similarly, ül tells stories that emerge from everyday and historical experience of the Mapuche in their community territories and more generically in life. In corridos or ül many of the told stories are also tied to local history, and therefore contribute to collective memory-making rooted in local or specific social, political, and historical events.

A second common pattern is the sentimentality and emotional drive that imbued in the corporeal, vocal, and thematic expressiveness of rancheras, rap, and ül. To some extent, what Mapuche scholar Ramón Curivil Paillavil has stated regarding ül can be applied to these genres: They "seek to reach the heart" (Curivil Paillavil 2002, 23). Even, in the case of corridos, the recourse of the *voz lastimera* evidences this in the vocality of the music, as also happens in the emotionally affected vocalization and the commotion of both the chanter and the listener that the aural and musical experience of certain types of ül entails. Undoubtedly, this emotional vibration is at the heart of rap music, too, to express collective or subjective feelings, which can range from rage or sorrow to joy. Finally, types of ül like ayekan-ül can converge with certain rap songs or rancheras as musical forms aimed to animate and uplift the spirit of audiences, as exemplified in songs we have discussed in this chapter, such as "Celinda" (ül) by Elisa Avendaño Curaqueo and "Mapudunfinge" by Wechekeche Ñi Trawün. This emotional dimension traverses the variety of musical registers, sounds, and listening practices of Ngulu Mapu.[41]

Contemporary Mapuche music constitutes a multilayered space and temporality. The sentimental and rhythmic attachment to Mexican rancheras and corridos, rap, and ül all speak of a sentient, communicative, and cognitive Mapuche engagement with music. Undoubtedly ül provides a grounding dimension in Mapuche music as a long-standing tradition native to Wallmapu, as the Mapuche chant that sonically connects us as a people of long history as a music rooted in our own rakiduam (diverse thinking) and kimün (knowledge system). However, as we have heard throughout this chapter, it is a music that also coexists with other auralities and vocalities in the Ngulu Mapu, a territory ultimately rooted in what we call Itro Fill Mongen—an ecosystem diverse in lives and voices.

Coda

Tralcao is a significant territorial and aural reference in the trajectory of my life. It is the name of my tuwün *(place of origin), in the north part of the Mapuche-Williche territory of Ngulu Mapu. As I have pointed out in autobiographical vignettes, I spent my childhood and part of my adolescence in this rural area of the Füta Willi Mapu, from 1960s to the 1980s. It is where I learned to listen, with my ear burrowed into the sonic environment of a colonized Mapuche-Williche community. It is where I experienced the enveloping waves of acoustic colonialism as well as the volatile and episodic forms of Indigenous interference that subsisted in those colonized lands of Wallmapu.*

....

TRALCAO—HUECHANTE

As indicated at the opening of this book, Tralcao derives from the noun *tralkan* (thunderstorm).[1] It may originally have been *Tralkawe*, "place of thunderstorm." As a Mapudungun-rooted term distorted into Spanish, Tralcao is an example of a specific operation that settler colonial discourse inflicts on native voices and sounds. This "acoustic disfigurement" underscores colonial mediation in the very grammar and sound of language, an effect of the broader dominance of Spanish phonetics in Ngulu Mapu. In the case of a Mapuche-Williche community such as Tralcao, it reflects the reality of "linguicide" (Derince 2013; Turin 2012; Charny 1994), a key dimension of acoustic colonialism. Additionally, as a native toponym, Tralcao illustrates what scholar Lorenzo Veracini calls "name confiscation" (2010, 47–48), which occurs when settler states keep Native toponomies and place labels that lend a degree of "indigeneity" to a region. Mapudungun-rooted names abound throughout central and southern Chile, usually inscribed under the mediation of colonial grammar and phonetics.

Nevertheless, from a Mapuche aural position, toponyms such as Tralcao also contain resonances that connect us to a sense memory of the place as belonging to Wallmapu—to *not-Chile*—as well as a mnemonic trace of the sonic dimensions of the land. I recall how lightning and thunderstorms shook the walls and windows of my grandmother's house. I learned to engage the earth's sonorities—the acoustics of the forest, the unsettling thunder, the hissing wind, and the rain on the zinc roof. Thunderstorms were the heavy metal music of the Füta Willi Mapu. Thus, Tralkawe became my school for the multitonal music of the Mapu. I learned that each native territory has its enduring sense and sound, despite the entangling colonizing powers.

Tralcao was founded during the second half of the nineteenth century as a reducción, mostly populated by members of the Huechante and Pangui kinship. At the time, Mapuche-Williche communities in the north of what is today the province of Valdivia feared a possible Chilean military invasion. From the 1860s to the 1880s, the so-called Pacification Campaign had resulted in the forced removal of many Mapuche communities in the nearby Araucanía region (Correa Cabrera 2021, 125–226; Mariman Quemenado 2006, 101–13). Amid this environment of fear, Pedro Nolasco Huechante, a longko (community head) of what had been the Lof Huechante, requested "protection" from the Capuchin mission in the region.[2] By "donating" vast extensions of land to the mission, the community got the needed protection and resettled in a smaller territory. The reducción of Tralcao became a refugee zone during those decades of intensive colonization in Ngulu Mapu.

The community had to compromise its Mapuche lifeways and submit to a colonizing process of Christianization and Chileanization. Spanish supplanted Mapudungun, Mapuche ceremonies disappeared from collective life, and the Capuchin mission and other settlers absorbed the vast territory of the Lof Huechante. Surrounded by rivers, creeks, and wetlands, Tralcao was established on a quasi-island. Over time, several fundos (large estates) rose on the edge of those water borders. Mostly owned by German settlers (Heise, Grob, Heinrich, Hoffman), they formed a ring around Tralcao. Later, the neoliberal "structural adjustment" of Chilean capitalism of the mid-1970s further undermined the Mapuche economy and campesino communities. All land, including native territory that had acquired some protection under the Agrarian Reform of the 1960s and early 1970s, was subjected to the logic of the free market (Levil Chicahual 2006, 228–34). In the 1980s, Mapuche-Williche from Tralcao started migrating to cities. By the early twenty-first century, many Williche families had sold their lands to non-Indigenous people from neighboring cities. This increased the "parcellation" of the territory and converted it into a village, different from the predominantly Williche Tralcao.

Previously, the sonorities of native surnames such as Pangui, Millanao, Lefno, Caurepan, Martin, Huincatripai, Agregan, and Huechante shaped the aurality of the community. There was ample space for seeing, hearing, and feeling a Mapuche-Williche territoriality in Tralcao. In 1974, to complete my elementary education, I moved away to Pelchuquin, a small town dominated by settler Creoles and German families where the Catholic mission and the settler-owned fundos were omnipresent. Since the public school was for children from low-income families, which was the case of most of Mapuche homes, there I met classmates from Laken Mapu (Lands of the Coast), besides many Williche peers.

In 1976, I experienced a much more drastic change in my aural and racial environment. To attend high school, I had to move to Valdivia, the capital of the province whose name honors the Spanish conquistador, Pedro de Valdivia. Among my classmates, surnames rooted in the kinships of our Füta Willi Mapu, other than mine, were no longer present. The White colonial and racial dominance was clear when teachers took roll. In the conspicuous aural presence of Hispanic, German, Italian, and Yugoslavian surnames, *Huechante* sounded dissonant. At that point of my life I lacked full consciousness of my Mapuche-Williche belonging, so hearing *Huechante* read aloud made me feel uncomfortable, or at times quietly embarrassed, especially on occasions when *Huechante* was awkwardly pronounced and murmurs and discreet laughter arose from the back of the classroom.

These episodes led to my developing unconscious "shame" about my background, an experience familiar to very many Mapuche in the anti-Indian racial setting of "modern" Chilean society.[3] Indeed, this "educational" stage of my life signaled my definite immersion in the hegemonic milieu of settler Chilean society. Yet it was also a phase in a long and contradictory process of becoming aware of the role of language, naming, and phonetics in the deployment of acoustic colonialism. These experiences permeate the roots and routes of the formulation of the acoustic and sonic inflections of this book.

UNDER THE NOISES OF ACOUSTIC COLONIALISM: MAPUCHE INTERFERENCES

Epistemologically and methodologically, this book has aimed to put in practice the Mapuche concept of *allkütun*, "to listen attentively," as a mode of engaging literary, radiophonic, and musical texts. This practice is permeated by and cross-fertilized with the knowledge I acquired in my formative experiences of listening and "sound relations" (Perea 2021), particularly in the colonized Mapuche-Williche environment in which I grew up. As an epistemological, methodological, and corporeal concept, allkütun has a critical singularity anchored in centuries of Indigenous thought rooted in the lands of Wallmapu; therefore, in the aural impulse of this book, it is a key notion to critically advance ways to think of, imagine, and/or put in practice what has been called "the decolonization of the ear" (Denning 2015, 137). Additionally, my approach has been enriched in substantive ways by recent theoretical and methodological approaches from interdisciplinary sound studies in connection with questions of colonialism and indigeneity (Perea 2021; Ochoa Gautier 2014; T. G. Reed 2019; Robinson 2020); and from critical studies of race and racialization (Bronfman 2016; Campt 2017; Eidsheim 2015; A. Reed 2021; Stoever 2016).

The first two chapters analyze settler colonial representations of the Mapuche as the result of the mediation of the colonial ear, a cornerstone in the articulation of acoustic colonialism. What emerges is a colonial and racialized politics of acoustic disfigurement on Indigenous subjects. Chapter 1 is a critical, attentive listening of Chilean writer Alberto Blest Gana's novella *Mariluán* (1862). Chapter 2 delves into the vocal and musical performance of Indio Pije, an "Indio" personage performed by Chilean comedian and singer Ernesto Ruiz in the radio comedy *Residencial La Pichanga* during the 1960s and 1970s, which led to a series of cumbias in the mid-1970s. Vocal and sonic modes of caricaturing and misrepresenting the Mapuche were conspicuous traits in the literature of Blest Gana and the vocalization of Ruiz's Indio Pije, adding a racialized layer to a primitivizing portrayal of indigeneity in mainstream Chilean society. These literary, musical, and radiophonic practices were part of the acoustic domination of Ngulu Mapu articulated through the colonial ear: a corporeal, perceptual, and symbolic membrane that filters and demarcates the figuration, or rather the disfigurement, of the colonized human and nonhuman territory. This gives shape to a long-standing media regime of sound and listening that erode Indigenous subjectivity and collectivity, resulting in acoustic distortion, stigmatization, and historical erasure of the Mapuche.

Sound becomes a field of struggle replete with frictions and clashes between colonizing powers, human and nonhuman forces. In this sonic battle, Mapuche interferences take shape as "enduring indigeneity" (Kauanui 2016)—that is, acts that not only respond but also generate vocalities and sonorities alternative to acoustic colonialism. Thus, chapter 3 approaches the role of poetry in projecting Mapuche vocal and sonic practices. The creative works of Lorenzo Aillapan Cayuleo and Leonel Lienlaf mutate into onomatopoeic utterances or chants. The Mapu—land, earth, territory, space, universe—emits its whispers, screams, or elliptical silences. Chapter 4 analyzes the Mapuche use of radio as a tool to build sonic and aural autonomy. Completing this journey, chapter 5 is an overview of the musical palimpsest that forms the aurality of Ngulu Mapu, the culture of listening to Mexican popular music, the more recent cultivation of homegrown rap, and the persistent tradition of *ül*.

Indigenous interferences are immersed in the broader territorial and political movement to dismantle the settler colonial, capitalist, and patriarchal nation-state powers that subjugate the life of Ngulu Mapu and Wallmapu. As cultivated and used by the Mapuche, the aural and sonic territories of poetry, radio, and music compel us to expand our notion of the political. They constitute liberatory acts and arts involving language, bodies, voices, and sounds that, tenuously or emphatically, put in action forms of sonic autonomy and disrupt the acoustic waves of hegemonic powers. Indigenous interferences therefore entail acts that invite us to imagine ways to challenge centuries of colonialism, vocally and sonorously.

FROM INDIGENOUS INTERFERENCE TO DREAMING A FREE WALLMAPU

Indigenous interferences should be heard as disruptive and critical forces that cannot be accommodated within the script of settler colonial and capitalist state powers. The Mapuche People never had a centralized authority, a sovereign, or an imperial ruler. Instead, the social and political formation has consistently been polycephalic. Unlike other Indigenous and non-Indigenous civilizations, we did not adopt an imperial form of rule, nor did we have a central government along the lines of what wingka "modern" societies regard as the state, for state, empire, and even "nation" come from Western political and philosophical imaginaries. We were always many territories with local forms of social life and structures of governance (Lof, Rewe, Wichan Mapu, Meli Wichan Mapu). Our local communal authorities gathered to make agreements and acknowledge disagreements across different territories. We have been many peoples within one people, the Mapu Che.

Colonialism across centuries, from Spanish to Chilean and Argentine colonial powers, along with its attendant capitalism, destroyed our cultural, social, and political life system. Nevertheless, for many Mapuche communities, this political and historical memory of being a multiterritorial society remains part of our sense of governance and life, for today and tomorrow. Our history as a Mapuche society, therefore, is anticolonial, anticentralist, and anti-imperial. It emerges not from an abstract "future" but from the historical memory left to us by our *kuifikeche* (ancestors). Free Wallmapu, in this sense, does not find an equivalence in the placeless notion of "utopia" or "futurism" embraced by Western avant-garde ideologies of the past century. Free Wallmapu is not a utopia, because it already *took place* in history. To question present oppressive systems and to imagine our way toward tomorrow, the Mapuche turn our historical gaze and listening toward what is temporally behind us interwoven with what is upcoming.

The aural environments of Ngulu Mapu and, more broadly speaking, of Wallmapu, are of a people who possess multiple voices and sounds. The aspiration for a Free Wallmapu should be imagined as the reconstruction of a plurivocal and polyphonic environment. The human and nonhuman forces that resist colonial oppression in Ngulu Mapu and Puel Mapu have always been here and there, as voices, calls, chants, or fighting cries. "Multicultural" or "intercultural" politics of colonial nation-state assimilationism cannot contain the plurivocal, polyphonic, and multiterritorial nature of Mapuche society. Furthermore, these dimensions are tied to a deeper sense of life's plurality, embodied in our ethos of land relations. As a "People of the Earth," our human imaginaries follow the logic of the Mapu to whom we the Che belong. That is the principle of Itro Fill Mongen, which defines the Mapu as an ecosystem of multiple, diverse lives. Free Wallmapu is therefore an impulse toward the liberation of the acoustic and sonic multiplicity that comprises the Mapu. To feel, to think, to listen, and to imagine through the horizon of a Free Wallmapu entails a sense of liberation away from the monocephalic and single-stringed script of colonial, capitalist, and patriarchal powers.

I conclude by invoking this broader aspiration for liberation because it is not exclusive to the Mapuche People but is shared by the oppressed peoples, lands, and waters who struggle for life across regions and continents amid a global ecological crisis. I invoke this call for the recovery of a plurivocal and polyphonic sense of community with a personal sense of hope and with my ears open to my Mapuche-Williche assignment in life. The marker of that call is suggested in my maternal last name, Huechante: *Weche* (new, young) and *Antü* (day, sun). Weche Antü: new day, new sun.[4] To honor its semantics, day to day I embrace the hope for the advent of another time. That aligns my heart and my mind with sectors of the Mapuche movement that embrace a decolonizing politics of autonomy and advocate for the Itro Fill Mongen and the liberation of the Mapu. This historical struggle takes place in times of colonial, capitalist, and patriarchal domination—or what Mapuche-Tehuelche writer and leader Moira Millán characterizes as *terricidio* (2024). This book is an attempt to unveil one dimension of those oppressive powers—acoustic colonialism—while longing for a new sun or new day in which the multiple territories, lives, and voices of the Mapu not only interfere but rise and liberate.

NOTES

INTRODUCTION

1 Around 1544, in the name of Spanish conquistador Pedro de Valdivia, his envoy Gerónimo de Alderete "took possession" of what today is this urban center. Pedro de Valdivia named it Santa María la Blanca de Valdivia in 1552. Later, the city became known simply as Valdivia. Before the Spanish colonial settlement, it was an important epicenter of Mapuche-Williche and Mapuche-Lafkenche life around the main river of the area. According to the Spanish chronicler Alonso de Góngora Marmolejo, the native name of this place was *Ainil*. This term seems to be Góngora Marmolejo's Castilianization of the Mapudungun term *añil*, which refers to the black mud formed in the wetlands so common around the rivers of the Füta Willi Mapu. On Alderete's expedition in the region and the Spanish arrival in what would become Valdivia, see Góngora Marmolejo [circa 1576] 1862, 224.

2 In Mapudungun, *mariküga* means "ten lineages."

3 During the period of Spanish colonialism, as well as prior to the 1840s—that is, during the foundational decades of the Chilean nation-state—Jesuit missionaries played a major role in the religious colonization of Mapuche territories.

4 The Capuchin Order is a Catholic order of Franciscan friars. The first Capuchin missionaries arrived in Ngulu Mapu, or what is today known as southern Chile, around 1848. Capuchins were instrumental in establishing Catholic missions in several areas of what is today mapped as the Provinces of Cautín, Valdivia, and Osorno, particularly in Mapuche territories belonging to the Nagche (People of the

Lowlands), Wenteche (People of the Highlands), Lafkenche (People of the Coast), and Williche (People of the South). During the Capuchin missionary era, Tralcao, my Mapuche-Williche community of origin, became part of the radius of influence and control of the Mission of Pelchuquin, symbolically founded in 1860—that is, the same year the Pacification Campaign, the Chilean state-sponsored military invasion of Ngulu Mapu, was launched. On the Capuchin missions in Mapuche territory, see the firsthand account by the Capuchin priest Antonio de Reschio, originally published in Italian in 1890 (Reschio 2018). Mapuche-Lafkenche scholar Héctor Nahuelpan Moreno's doctoral dissertation also offers an insightful account on the role of the Capuchin missions in the colonization of Ngulu Mapu (see Nahuelpan Moreno 2013, 192–216).

5 Referring to the Alaska Native Peoples's experience of "missionary colonialism," Jessica Bissett Perea recounts how Quaker missionaries settled in Inuit territory in the early twentieth century. In the case of the Nuurvik community, as soon as the Quaker missionaries arrived there in 1914, they silenced the Inuits' native drumsongs practices. Stigmatizing them as expressions of "primitive idolatry," the missionaries banned Inuit music and dancing in the community (2021, 86). It is interesting to note how colonial logics of missionization operate in a similar way at both ends of the hemisphere.

6 Between 2009 and 2015, the Comunidad de Historia Mapuche published two multiauthored books, namely, *Ta iñ fijke xipa rakizuameluwün: Historia, colonialismo y resistencia desde el país Mapuche* (2012b), and *Awükan ka kuxankan zugu Wajmapu mew: Violencias coloniales en Wajmapu* (2015). These Mapuche authorial interventions reflected our collective aspiration to contribute to what Mapuche researcher Herson Huinca Piutrin, one of the CHM cofounders, conceives of as a critical dimension of "a process of decolonization" in the realm of written intellectual production. According to him, this process necessarily entails that "the Mapuche stop being [mere] objects" of study and exercise "sovereignty over the knowledge and reflection on our diverse society, from within" (Huinca Piutrin 2012, 117).

7 In the book *¡ . . . Escucha, winka . . . !*, Pablo Mariman Quemenado's chapter on the colonization of Mapuche territories by the Chilean and Argentine states has been extremely relevant in this historical discussion (Mariman Quemenado 2006). Also, the multiauthored introduction from our first book as Comunidad de Historia Mapuche was a significant impulse for the contemporary emergence of an anticolonial Mapuche thought (Comunidad de Historia Mapuche 2012a, 11–21).

8 In my view, Wolfe's theory of "settler colonialism" cannot be "applied" as the sole analytical angle to account for the heterogeneity of histories, experiences, and forms of colonialism in what is today known

as "Latin America." For an excellent discussion on Wolfe's theory, see Kauanui 2016; Shoemaker 2015.

9 Regarding the development of sound studies in Latin America vis-à-vis questions of colonial relations, it is worth mentioning here the work of researcher and artist Mayra Estévez Trujillo. In Quito, Ecuador, in 1996 she became a member of Centro Experimental Oído Salvaje, a sonic art and radio collective. She has published the bilingual book *Estudios sonoros: Desde la región andina / Sonic Studies: From the Andean Region* (2008). See Estévez Trujillo 2008.

10 According to scholar Mark Turin, Ukrainian Canadian linguist Jaroslav Bohdan Rudnyckyj coined the concept of "linguicide" in the 1960s "while exploring the fate of his native Ukrainian under Russian linguistic and political pressure" (1994, 849). Another author, Israel Charny, defines linguicide as "forbidding the use of or other intentional destruction of the language of another people," constituting "a specific dimension of ethnocide" (1994, 77). More recently, in an article on the status of the Kurdish language in Turkey, scholar Mehmet Şerif Derince relates linguicide to a state politics of "linguistic homogenization," which would "seek to aggrandize Turkish as the national language at the expense of killing other languages and squeezing the diverse body of its citizenry in a monolingual and mono-cultural straitjacket" (2013, 146). None of these authors links the term to the articulation of colonialism. As discussed in my study, the erasure or elimination of a native language—like the case of Mapudungun in the Mapuche-Williche context—plays a strategic role in the warping of acoustic colonialism and, more broadly speaking, in the establishment of settler powers on an Indigenous territory.

11 In its phonetics and meaning, the very denomination *América*, or *the Americas*, is a telling example of the acoustic and semantic colonization of this continent. In response, Indigenous leaders and scholars have proposed ways to avoid the colonial lexicon and restore Native modes of naming. To replace *América*, *the Americas*, or *Latin America*, one name that has been used is *Abiayala*, a term from the language of the Guna People. See Keme 2018.

12 *Criollo* is the ethnic and racial denomination of people of Hispanic or other European descent who were born in Latin America and became the dominant socio-racial sector under the rule of the post-1800 settler nation-states or republics. In this book, Creole(s) will be used as a translation for *criollo* or *criolla*, and for its plural forms. Criollo, in the context of power relations in Chile and other Latin American societies, is also equivalent to the term *settler*, as understood in settler colonial analytics (Wolfe 2006; 1999).

13 Regarding the names of places and peoples in Mapudungun on Mariman Quemenado's maps, I refer to them as written by the author. It

seems that Mariman Quemenado's writing alternates between the two most influential grammars for Mapudungun: the Raguileo Alphabet and the Unified Alphabet.

14 *Chile,* as written in the Spanish original of the Inca Garcilaso de la Vega's *Comentarios reales.*

15 According to scholar José Manuel Zabala, the parlamento was a high-level meeting of representatives from the societies in conflict and the "main institution of Spanish-Mapuche negotiation on the frontier." It "emerges towards the end of the XVI century, developing and consolidating during the XVII century and becoming quite complex and formal throughout the XVIII century." He also adds that this type of negotiating summit "was used most broadly during the last quarter of the XVIII century and beginning of the XIX century" (18).

16 Mapuche economics of livestock mostly included cattle, equine, and ovine. In Mapudungun, the term *külliñ* originally referred to the livestock; but over time, because of the new colonial economic exchange system, it became a word for "money."

17 Horacio Lara was a Chilean military officer and writer born in the city of Concepción in 1860. In 1887, he became an official in the Chilean army. Lara participated in the last phase of the military invasion of Ngulu Mapu. He published his *Crónica de la Araucanía* in two volumes, in 1888 and 1889. Jorge Pinto Rodríguez argues that Lara—along with Leandro Navarro, another army officer from the same period, and the scholar Tomás Guevara—belongs to the trend of "regional historians" who, unlike other nineteenth-century Chilean historians, do not erase the Mapuche experience from their writings (2003, 246–51).

18 For a critical overview of state terrorism in "intercultural" or "multicultural" neoliberal Chile, see Richards 2013, especially her chapter titled "Constructing Neoliberal Multiculturalism in Chile" (101–33).

19 On the mid-1970s monetarist policies and market reform of Chile's so-called structural adjustment, see Klein 2007; Moulián 1997; and Valdés 1995. On the cultural and symbolic effects of this process in Chilean society, see Cárcamo-Huechante 2007, 2006.

20 Voiced at minute 1:40 of this video. This is my translation of these verses. For a more extensive scholarly analysis of Huichaqueo's short film, see Gómez-Barris 2017, 66–90.

21 Here I use the notion of apostrophe in its classical sense, as "a figure of speech in which a thing, a place, an abstract quality, an idea, a dead or absent person, is addressed as if present and capable of understanding" (Cuddon 1998, 51).

22 On literary scholarship that brings together "phonemic analysis" and close reading, see Stewart 2002.

23 On the status and relevance of "fieldwork" in the humanities, see Castillo and Puri 2016, 1–26.

24 In addition to Dylan Robinson's conceptualization of "critical listening positionalities" (2020, 9–11), we can find similar theorizations on listening as a method in other sound studies works. For example, French critic and filmmaker Michel Chion has coined the concept of "reduced listening" to describe a "listening mode that focuses on the traits of the sound itself" in audiovisual media ([1990] 1994, 29). In musicology and cultural studies, Colombian scholar Ana María Ochoa Gautier has delved into the domain of the aural in the archive and "listening" as "simultaneously a physiological, a sensorial, and interpretive cultural practice" (2014, 25). In a similar vein, Cuban American scholar Alexandra Vazquez has worked on the suggestive methodological notion of "listening in detail" (2013, 17–20). Equally important for my own approach are the elaborations of American studies scholar Jennifer Lynn Stoever on "the aurality of race and the unspoken power of racialized listening" (2016, 7). Other authors who have elaborated on critical modes of listening are Campt (2017); Eidsheim (2015); Furlonge (2018); Kheshti (2015); Marsilli-Vargas (2022); Nancy (2002); Vazquez (2013); and Voegelin (2010).

25 Essayist Peter Szendy returns to a classical source, the *Dictionnaire de l'Académie française* of 1694, where the verb *écouter* is defined as "oüir avec attention, prester l'oreille pour oüir" (2007, 24).

CHAPTER ONE. DISFIGURING AND SILENCING OF THE MAPUCHE IN THE 1860S

1 In 1861, Colonel Cornelio Saavedra presented to the Chilean government a plan for the "pacification" of those territories still under Mapuche control, mostly in the Bío-Bío and Araucanía regions. This plan called for erecting a line of fortification around the Malleco River to expand the Chilean state's geopolitical control from its boundary at the Bío-Bío River. As a result, by 1882 the Chilean state had gained military and administrative control of the Araucanía region.

2 Manuel José Olascoaga (1835–1911) was a military man, explorer, writer, painter, and engineer originally from Mendoza, Argentina. Between 1869 and 1871, he joined Colonel Cornelio Saavedra during the campaign of Chilean settler occupation of Ngulu Mapu. In Argentina, Olascoaga joined General Julio Roca in the conception and implementation of the so-called Conquest of the Desert, the military invasion of Puel Mapu.

3 As mentioned at the start of this chapter, Olascoaga's painting is a visual image that has become familiar for Indigenous and non-Indigenous

readers in Chile through school textbooks and, more recently, digital media. It has been reproduced in illustrated books of influential Chilean historians, such as Francisco Encina's *Resumen de la historia de Chile* (1954), and Sergio Villalobos's *A Short History of Chile* (1996)—first published in Spanish as *Breve historia de Chile* (1983). Villalobos has become a firm and polemical/outspoken advocate of the legitimacy of the colonial settlement of the Chilean nation-state on Mapuche lands. I am indebted to the work of British scholar Joanna Crow, which prompted me to think about my own exposure to this already canonical image in Chilean visual and historiographical culture. In her book *The Mapuche in Modern Chile: A Cultural History*, Crow keenly comments on this painting and the influence it has attained through dissemination in school textbooks (2013, 26–28; 237). As an example, she highlights Verónica Méndez Montero, Carolina Santelices Ariztía, Rodrigo Martínez Iturriaga, and Isidora Puga Serrano's school textbook *Historia, geografía y ciencias sociales. Texto para el estudiante* (2009), published by the educational publisher Santillana, and a 2008 edition of Villalobos's cited book, the circulation for each of which has been widespread.

4 In *Resumen de la historia de Chile Tomo II* (1954), historian Francisco Encina describes Olascoaga's painting as an image of a "reunión amistosa" (1954, 1287). The same description is replicated in Villalobos's *Breve historia de Chile* (1983, 141). To further contextualize, I should add that throughout the so-called Pacification Campaign, the communities of Ngulu Mapu adopted divergent positions on how to respond to Chilean settler agents at certain moments. According to existing historical accounts, some communities of the Nagche (Mapuche of the Plains) and Wenteche (Mapuche of the Upper Lands) opposed and confronted the invaders, while sectors of Lafkenche (Mapuche of the Coast) and Pewenche (Mapuche of the Cordillera) were forced to yield before military pressure and negotiate. The Hipinco Parliament is an example of this latter scenario, as a meeting that took place during what historians have characterized as the second phase of the Chilean "Campaign" (between 1867 and 1869), a period of intensification of colonial military violence in Mapuche territories. For further problematization of the "story of peace and friendship" fabricated by Chilean settlers, see Crow 2013, 26–30.

5 For a literary, cultural, and historical view on the Campaign or Conquest of the Desert in post-1979 Argentina, see Viñas [1982] 2003, 17–27, 53–72.

6 Cited in Viñas [1982] 2003, 61.

7 On the specific idea of the gaucho as the "hombre argentino," see Ludmer [1988] 2000, 42–44.

8 The newspaper *Aurora de Chile* published fifty-eight issues between February 20, 1812, and April 1, 1813. As the first newspaper in the country, it became a print symbol of the newly formed settler colonial nation-state of Chile.

9 *La Voz de Chile* was published in Santiago between 1862 and 1864. It was founded by two members of the liberal Creole elite, brothers Guillermo and Manuel Antonio Matta, who were affiliated with the Liberal Party until 1857 and then with the foundation of the Radical Party. During its short life, *La Voz de Chile* published several works of prose fiction, following the conventions of the newspaper serial. *Mariluán*—along with two other narratives, *Drama en el campo* and *La venganza*—was part of series of three short novels published in *La Voz de Chile* in 1862; later the same year, the press of the newspaper published them together in a single book titled *Drama en el campo* (Drama in the Countryside). This trilogy has been republished under the same title on several subsequent occasions, including in 1876 in Paris and later, in 1949, in Santiago. In 2005, *Mariluán* under the title *Mariluán: Drama en el campo*, was printed in a separate edition by Lom Ediciones press in Santiago, edited by Chilean literary scholar Amado Lascar. All citations in this chapter are from the 2005 edition.

10 In Mapudungun, *Mariluán* means "ten *guanacos*" (*mari*: ten; *luan* [*guanaco*]: a native South American ruminant, *Lama guanicoe*, to which other native quadrupeds, such as the llama and the alpaca, are related). *Mari*, per se, is a highly relevant number in the Mapuche language, as evidenced in the standard way to say hello: "Mari mari." Mapuche names abound in human and nonhuman animal associations. *Mariluán* is thus a name that suggests relevance and strength. As I will note, Mapudungun is markedly absent from Alberto Blest Gana's novel; it only appears through onomastic discourse, namely, to name Mapuche characters (i.e., Mariluán, Cleu, Peuquilén).

11 Lonko Juan Francisco Mariluán was Mapuche-Wenteche. He was a leading tribal representative in the Parliament of Tapihue of 1825 in which Mapuche authorities were able to reach some diplomatic agreements with the newly formed "Republic of Chile," which was represented by Colonel Pedro Barnachea. In 1826, lonko Mariluán rebelled in a context of increasing tensions between the Chilean state formation and Mapuche people. See Marimán Quemenado 2006, 82–83.

12 For an early critical elaboration on Blest Gana's linkage to literary realism, see Latcham 1959(?). Guillermo Gotschlich has additionally analyzed the Chilean author's cultivation of the subgenre of the "historical novel"—in a Lukacsian sense. His discussion focuses on Blest Gana's novel *Durante la Reconquista* (1897) as an example of this path (Gotschlich 1991).

13 By the mid-nineteenth century, Talcahuano was an important naval and commercial port town in the region of Bío-Bío, southern Chile.

14 American scholar Doris Sommer praises the novel *Martín Rivas* as "being paradigmatic for national romances" (1991, 217). Sommer's celebratory approach to Blest Gana's fiction does not problematize the racial, colonial, and capitalist layers of Creole supremacy that underlie the very fabric of the Chilean nation and Blest Gana's "foundational fictions."

15 "Expansion'" has been a recurring term in Chilean historiography to designate the period between the early 1860s and early 1890s. For example, the already cited historian Sergio Villalobos defines it as the era of "La Expansión," situating it between 1861 and 1891. He explains that, at this time, "the general prosperity, the population growth, and the need to increase agricultural production" in Chile "sparked a move to settle regions that had not yet been occupied" (Villalobos 1996, 138).

16 Also cited by Waldo Rojas (1997, 9).

17 In his essay, published in France in 1997, Rojas calls attention to the disjunction between "the representation of the heroized and ideal Araucanian and the flesh-and-blood Indian" that underlies this canonization of an epic, archetypical Native past (7). In a similar vein, Bernardo Subercaseaux's study offers a suggestive view of the reception of *La Araucana* in Chile, which he considers part of a "double discourse" that, on the one hand, glorifies the "Araucanians" of the past, following the discursive footprints of the Spanish poet, and, on the other hand, articulates a problematic assimilationist view of the "real" Indigenous subject. With a focus on the interpretation and reception of Alonso de Ercilla y Zúñiga's epic poem, Subercaseaux's article "La Araucana: Un texto que genera contexto" (2021) provides examples that range from the Araucanist discourse of early nineteenth-century Chilean state publications and the republicanist and independentist views of Creole leaders influenced by Ercilla y Zúñiga, such as Bernardo O'Higgins and Francisco Antonio Pinto, to Chilean Creole writers and foreign travelers of the 1840s and 1850s.

18 Another good example of the institutional status of *La Araucana* in Chilean state culture is its inclusion in the Biblioteca del Bicentenario, a book series created and sponsored by the governmental commission that led and oversaw the commemoration plans for the bicentennial of the "Republic of Chile" (1810–2010). In 2004, the Biblioteca del Bicentenario series issued an edition of Ercilla y Zúñiga's work under the title *Para gozar La Araucana de Alonso de Ercilla y Zúñiga Precursor de Chile*. See Ercilla y Zúñiga 2004.

19 *Malón* is a Mapuche term that refers to an "irrupción o ataque inesperado de indígenas" (Real Academia de la Lengua Española 2001, 1426).

In settler narratives of the nineteenth century in Argentina, Chile, and Paraguay, *malón* events frequently reference the capture of White women by male Native leaders.

20 The *kull kull* is a wind instrument made of a cow horn with a mouthpiece at its tip. On its characteristics and functions, see Pérez de Arce (2007) 2020, 341–52.

21 Regarding the representation of the Mapuche as "barbarians" or "savages," in his *Breve historia de Chile* (2018) scholar Alfredo Sepúlveda quotes a couple of telling examples of what prevailed as public discourse across the conservative/liberal divide in nineteenth-century Chile. From what has been called "the first conservative decade," he cites a speech of President José Joaquín Prieto before the Senate in 1835. Prieto, who led conservative governments between 1831 and 1841, states, "It should be kept in mind that the principles that rule among civilized nations are not applicable in the case of barbarians," to justify the capture of Mapuche children for domestic service in elite Creole homes during this period (Sepúlveda 2018, 181). On other side of the ideological spectrum, Sepúlveda cites liberal scholar Diego Barros Arana, considered the most important Chilean historian in the nineteenth century. By the end of the same century—in his *Historia general* (1891)—describing the motivations of the Mapuche people to join the Spanish royalist forces in the port of Talcahuano in 1817, Barros Arana referred to the Mapuche as "turbulent and rapacious savages" (Sepúlveda 2018, 550–51).

22 As demonstrated by scholar Francine Masiello in her study of gender and national discourse in nineteenth-century Argentina, the hegemonic discourse of the nation recurrently ends "proclaiming the triumph of the civilizing cause over the barbaric 'other'" (Masiello 1992, 9), the latter often being women, Indigenous peoples, or "other" oppressed subjects. Bounded within the civilization/savagery framework of the settler colonial and patriarchal Chilean nation, *Mariluán*'s novelistic ending follows this pattern. Indians and women end up being the "barbaric other."

CHAPTER TWO. INDIO PIJE

1 In Mapudungun grammars, it is written as *ruka*.

2 In their monograph on the history of radio in Chile, Lasagni et al. highlight radio communication's early emergence in Chile, the first transmission there occurring two years after the first one in the United States, almost at the same time as a similar inaugural emission in Argentina, and when only seven radio stations were operating in Europe (1985, 5). As historian Alejandra Bronfman notes on the

emergence of radio in the Caribbean, "The New York–based station WEAF opened sister stations in Havana in 1922 and San Juan in 1923" (2016, 5).

3 Radio scholar Margarita Pastene notes that by the end of the 1920s, there were fifteen radio stations in Chile. According to her, this medium experimented a substantial expansion in the country toward the 1940s, which was reflected in the fact that forty-nine stations got concessions from the state at this time (Pastene 2007, 115–16).

4 Prior to the arrival of radio in Chile, the phonograph was an important sound reproduction technology for conserving and reproducing human voices. Brought from the United States, the first phonograph was exhibited in the Chilean cities of Santiago, Valparaíso, Chillán, and Concepción in 1892. By the end of the nineteenth century, some phonographs were available for purchase in a store located on the Matte Passage in downtown Santiago, Chile. As in many other regions of the world, ethnographers of the time used this technology to record "Indian" voices, especially songs and stories. In 1907, Charles W. Ferlong used a phonograph to record songs by Selk'nam and Yahgan Natives in Tierra del Fuego, in the Patagonia region. Given the scarcity of the technology in Chile at the time, only a few ethnographers who traveled to Ngulu Mapu in this period, and who happened to be Catholic missionaries, had access to it. In the summer of 1910, missionary and surgeon of Jewish German origin Félix José de Augusta (born as August Stephan Kathan in Germany in 1860) traveled with a phonograph to the Mission of Panguipulli, located in the northeast of what is today Valdivia Province. With the collaboration of Friar Sigifredo de Fraunhaeusl, who was in charge of the mission, de Augusta carried out "the phonographic impression of some songs and an excerpt from a Mapuche parliament" (de Augusta 1934, iii). Unfortunately, there is no sound archive of phonograph recordings from this period and in the Mapuche context.

5 In 1932, Santiago-based station Radio Hucke started to broadcast, in Spanish, Italian playwright Dario Nicodemi's play *La nemica* (1924), with the remarkable voices of actress Maruja Cifuentes and actor Carlos Justiniano. See Pastene 2007, 120.

6 By the late 1950s, Residencial La Pichanga was housed in Radio Agricultura from Santiago, one of the major radio stations in Chile.

7 In 1965, Green Cross merged with Deportes Temuco to become Green Cross–Temuco; by the mid-1980s, it became Club de Deportes Temuco. In 2000, it adopted a new mascot, the "Lion of Ñielol," which was in use until 2002. During the following decade, Indio Pije was progressively revived as the team's mascot.

8 In the United States, the notion of "redface" resembles the practice of blackface, which consists of the performance "of white men ap-

plying and employing prosthetics to perform as black characters." It dates back to the American theater of the eighteenth century and was continued into the early twentieth century through modes and genres of representation such as "blackface minstrelsy" (Thompson 2021, 20–22). Examples of blackface even occur in other art forms and media in late twentieth and early twenty-first centuries.

9 On the uses of Indian mascots and racialized stereotypes, see Strong 2004, 81.

10 For example, in the early twentieth-century context, Philip J. Deloria mentions the Woodcraft Indians, Boy Scouts, and Campfire Girls as organizations invested in "the preservation of an allegedly disappearing native culture" (1998, 135).

11 The Santiago-based soccer club Green Cross was one of the nine inaugural teams that formed the first division of professional soccer in Chile from 1933 onward. Green Cross was founded in 1916.

12 The *kultrun* is commonly defined by musicologists as a Mapuche percussion "instrument," akin to a drum or timbale. Yet for the Mapuche, a kultrun is more than that, embodying the whole Mapuche universe and multiple dimensions of our life relations as Mapuche, that is, "People of the Earth." The kultrun is made from a hollowed tree trunk and covered with a stretched horse, sheep, or cow hide. For an explanation of the characteristics and functions of the kultrun in musicological terms, see Pérez de Arce (2007) 2020, 144–65. For a Mapuche view on it, see the documentary video *Aukinkoi ñi Vlkantun: Hace eco mi música* (Avendaño Curaqueo 2010).

13 Scholar Manuel Alcides Jofré points out that throughout Chile, during the twentieth century, "a national network of kiosks" in cities, which were "assisted by a set of distribution agencies," became critical to the wide circulation of revistas de historietas, magazines, and newspapers. He states that in this period, kiosks existed "at all social levels, across all diverse economic landscapes"; and when there were no kiosks available within an area, newspaper peddlers used to cover revista sales (Jofré 1983, 42).

14 Chilean populist politics had an important antecedent in the right-wing leadership of Arturo Alessandri in the early twentieth century, especially during his first presidency (1920–24).

15 Describing the changes of the 1950s and early 1960s, historians Simon Collier and William Sater write, "The old 'print culture' was now challenged by the advance of the electronic media, though not supplanted: more than 4 million books (1,400 separate titles) were printed in the Chile of 1959, and a visitor that year would have found a wide variety of magazines (political, humorous, sporting, feminine, right-wing, left-wing, Catholic, masonic) and newspapers available at the traditional sidewalk kiosks (*quioscos*) in the cities" (1996, 295).

16 One of Chile's first professional soccer teams, Magallanes, was named after the settler figure of Ferdinand Magellan, thus, like Green Cross, maintaining Chilean soccer's attachment to the sport's European roots.

17 In 1962, the Venezuelan singer Luisín Landáez arrived in Arica, northern Chile, and in 1964, Colombian artist Amparito Jiménez settled in Santiago. Both became iconic voices that helped cumbia achieve its position in Chilean mass culture. The most popular cumbia band in the country, Sonora Palacios was founded in 1962, paving the way for the proliferation of electronic-instrument-based cumbia, or what became defined as "sonorous-style" cumbia. As of today the most comprehensive study of cumbia in Chile is *¡Hagan un trencito! Siguiendo los pasos de la memoria cumbianchera en Chile (1949–1989)*, a book authored and edited by Lorena Ardito, Eileen Karmy, Antonia Mardones, and Alejandra Vargas (2016).

18 Citing a study conducted by the marketing department of the Catholic University of Chile's Centro de Investigaciones Económicas, César Albornoz notes that, by 1972, Chile—at the time a country of 8,884,768 people—had around 600,000 homes with a television, 400,000 of which were in Santiago and 200,000 in provinces (Albornoz 2014, 146–47). Compared to radio communications, television still had a limited reach, concentrated as it was in the capital, while radio extended throughout the country. In the 1960s, the medium reached the broader public as low-price portable radios hit the market (Lasagni et al. 1985, 16). By the early 1970s, radio became the most popular medium in Chile, with local radio stations in all regions—from Arica in the north to Patagonia in the far south—and the wide coverage that some Santiago-based stations had at this time, with powerful frequencies on AM and shortwave, especially at night.

19 For a comprehensive review of the role of CORVI in Chile, see Aguirre and Rabí 2009.

CHAPTER THREE. LISTENING POETICALLY

1 In this period, besides Lienlaf's first book, several other books of poetry by Mapuche authors were published in Chile: *El país de la memoria* (1988), *El invierno, su imagen y otros poemas azules* (1991), and *De sueños y contrasueños azules* (1994) by Elicura Chihuailaf; *Profecía en blanco y negro* (1998) by César Millahuique; and *Ceremonias* (1999) by Jaime Huenún.

2 The Metropolitan Region refers to the greater Santiago area in central Chile. By the time the 2006 census was conducted, the Tenth Region (Region of the Lakes) included the provinces of Valdivia, Osorno,

Llanquihue, and Chiloé Island. In 2007, Valdivia Province became a new region, namely, Region of the Rivers.

3 The trope of pewma in Lienlaf's first book has been discussed in articles by Carrasco Muñoz (2002, 96–98) and Mabel García Barrera (2008, 36–37).

4 For an example of this type of approach, see Carrasco Muñoz 2002, 96–98.

5 Within *Se ha despertado el ave de mi corazón*, this poem is part of Lienlaf's textual series titled "Cholkiñmangey" / "Le sacaron la piel." See Lienlaf 1989, 30–37.

6 For an overview of the Chilean invasion and occupation of Mapuche territories from the mid- to late nineteenth centuries, see Bengoa 2000, 251–362; Correa Cabrera 2021, 25–352; Crow 2013, 19–36; Mariman Quemenado 2006, 101–22; Pinto Rodríguez 2003, 235–46.

7 As occurs with Mapuche names and last names, Lautaro was *castellanizado* (Hispanicized, or Castilianized). In Chile, his *castellanización* follows the linguistic trail of Alonso de Ercilla y Zúñiga's *La Araucana* (1569–89), in which Lautaro is often mentioned as a key figure. Lautaro has roots in two terms in Mapudungun: *lef*, which means swift, fast; and *traru*, hawk. Across decades, Mapudungun grammars elaborated by Mapuche and non-Mapuche scholars have come up with different ways to write it, such as Leftraru (Unified Alphabet), or Lvftraru (Raguileo Alphabet). In the first edition of *Se ha despertado el ave de mi corazón*, Lienlaf follows his own writing choice for it, namely, "Lautraro"; in a later edition, he resorts to the Raguileo Alphabet.

8 This return of Lautaro as an inspiring figure of resistance is present in different expressions of the late twentieth- and early twenty-first-century Mapuche movement. For example, Mapuche historian Fernando Pairican Padilla observes that, since the late 1990s, "the tactics of Toki Leftraru" are omnipresent in the combative actions for Mapuche "self-determination" and "national liberation" carried out by organizations such as the Coordinadora Arauco-Malleco (CAM), particularly by positioning the idea of the Mapuche activist as a *weychafe* (warrior). See Pairican Padilla 2015, 301–5.

9 This Mapuche-Lafkenche area is located around Lake Budi and the Pacific coast, toward the western side of the Province of Cautín; and its main town is Puerto Saavedra—a clear marker of Chilean settler colonialism in the region. Indeed, this town was named as such in 1906 as "a recognition of the State of Chile to General Cornelio Saavedra for his role in the occupation of the Mapuche territory" (Aillapan Cayuleo 2017, 22). Moreover, within Chilean geopolitics, this whole area is currently under the jurisdiction of what became equally named as Commune of Saavedra.

10 Ak'abal's poem "Cantos de pájaros II" is a good sample of an onomatopoeic practice close to what Aillapan Cayuleo does (Ak'abal 1995, 87). For similar examples, see Ak'abal (1990) 2008, 25, 35; Ak'abal 1995, 63, 87, 89.

11 George Steiner states, "If speaking man has made of the animal his mute servant or enemy—the beasts of the field and forest no longer understand our words when we cry for help—man's control of the word has also hammered at the door of the gods." Furthermore, the poet "is he who guards and multiplies the vital force of speech": "The poet makes in dangerous similitude to the gods" (Steiner 1967, 37). In this anthropocentric and patriarchal Western view, Steiner envisions the human and the poetic as quests for transcendence and elevation that places them above the nonhuman animal and earthly realm.

12 In Chile, another early twentieth-century literary work that undertakes a similar bird-flight imaginary, recalling the Western myth of Icarus, is the poetic novel *Alsino* (1920) by Pedro Prado.

13 In this book, I comment on the ceremony of Nguillatun, as well as other ceremonial events and religious authorities (i.e., Machi) in Mapuche life only within broad frames of reference, as long as they figure in texts or community contexts with which I am in dialogue. I limit myself, in this respect, to avoid endorsement of extractive and invasive ethnographical and anthropological approaches on Indigenous ceremonial and spiritual matters that are pervasive in an academy so accustomed to romanticizing, exoticizing, and/or primitivizing our cultural and religious practices and authorities as "ritual" or "shamanic."

14 In his own autobiographical narrative, Lorenzo Aillapan Cayuleo writes that Ngillatun "is the most solemn ceremony of Mapuche religiosity because many communities participate; eight of them organize it and are the ceremony's hosts, and another eight communities are invited. In the area where my family lives, it is prepared by eleven communities and another eleven are invited" (2017, 169).

CHAPTER FOUR. *WIXAGE ANAI*

1 In the epilogue of their seminal book, *¡ . . . Escucha, winka . . . !*, Mapuche historians Pablo Mariman Quemenado, Sergio Caniuqueo, José Millalen Paillal, and Rodrigo Levil Chicahual elaborate on how the Mapuche movement has interlinked the concepts of autonomy and self-determination since the 1980s and 1990s. "For us, autonomy represents an aspiration to recuperate a 'sovereignty suspended' by the invasion and conquest of the Chilean and Argentine State; this is a form of governance, a way to exercise our own administration in our territories." They add: "Autonomy is a tool for self-governance and for the exercise

of self-determination" (Mariman Quemenado et al. 2006, 253). It is worth mentioning here that the concept of self-determination has become increasingly prominent in twentieth-century international law and has been embraced by Indigenous movements worldwide as they ratchet up their demands for territorial rights and self-governance. According to S. James Anaya (1996, 75), self-determination constitutes "a universe of human rights precepts concerned broadly with peoples, including Indigenous peoples, and grounded in the idea that all are equally entitled to control their own destinies." For the historical, political, and juridical genealogy of this concept, see Anaya 1996, 75–96.

2 Among them, Ramón Curivil Paillavil, Clara Antinao Varas, María Catrileo, Elías Paillan Coñoepan, Margarita Elizabeth Huenchual Millaqueo, and José Paillal Huechuqueo were native speakers of the Mapuche language and fully bilingual (Mapudungun/Spanish).

3 The Jvfken Mapu Center was able to carry out and publish important collaborative and collective research. As examples of this work, it is worth mentioning their publications on the celebrations of the Wiñol Xipantu (the annual change to a new nature cycle) in Santiago (Malo Huencho et al. 2003), and on Mapuche identity and ceremonial practices in Santiago (Curivil Paillavil et al., 2001). Since these publications were published in the form of research documents, with a limited number of print copies, and were meant for circulation in the Mapuche community in Santiago, they are unique materials that are difficult to find in library archives.

4 The Jvfken Mapu Center always had the active participation of the *Wixage Anai* producers and broadcasters. But it also became a space for those who wanted to support *Wixage Anai* and collaborate in community research and fundraising initiatives. Notably, Carmen Malo Huencho, Ricardo Tapia Huenulef, Soledad Huaiquiñir, Fresia Paillal Huechuqueo, Olga Curinao, Víctor Pichun, and Emilia Cona were some of the Santiago-based Mapuche community members who participated in the center over the years.

5 Following sound studies scholar Jonathan Sterne, I regard "sensorial history" and "social" processes as interconnected (2003, 348). The intersection of radio, culture, and politics appears in several studies and critical interventions, starting with Frantz Fanon's writing on the role of radio in the anticolonial struggles for "national liberation" in French-occupied Algeria during the 1950s. Historian Alejandra Bronfman traces lines of intersection between questions of language (Creole, or Kreyól), race, and politics as staged through radio in Jamaica, Haiti, and Cuba in the 1940s and 1950s, as well as collective aspirations of decolonization and democratization and the process of forging public space for "sonic blackness" in the Caribbean (2016,

91–92). Her attention to "radio wars" or "radio battles" (Bronfman 2016, 140–43) resonates with my approach to the radiophonic medium in terms of acoustic colonialism and its countercurrents.

6 I acknowledge that the study of the *Wixage Anai* radio program's reception would be another highly relevant path of analysis to grasp its community impact and audience-formation process. This would entail a radically different approach, mostly rooted in sociology of media. Unfortunately, in part due to the complexity of mapping Indigenous audiences in Chile as well as the amount of human and financial resources that would be required for a sociological study of this nature, no scholar has pursued this line of investigation yet.

7 To some extent, this experience and process of building Mapuche agency resonates with a more recent elaboration of what Brandon LaBelle calls "sonic agency." Primarily, he defines it "as a means for enabling new conceptualizations of the public sphere and expressions of emancipatory practices" (LaBelle 2018, 4); moreover, in LaBelle's thought, "sonic agency" becomes "a support structure" for these expressions (2018, 21). Given that my own approach emphasizes the dimension of experience and process, instead of focusing on structures or organic counter-powers, I have preferred not to use LaBelle's notion of "sonic agency." However, if conceptually reframed, it could couple well with my own discussion on how a colonized, oppressed people build agency through sound.

8 In this period, Radio Nacional de Chile was housed in a three-story building on Lastarria Street near Alameda Avenue, within the historic Lastarria Barrio in downtown Santiago. Its AM frequency had enough power to cover greater Santiago during the day and all the regions of Chile at night; furthermore, listeners could tune in abroad thanks to shortwave coverage. As a state-owned broadcaster, Radio Nacional was strongly tied to the media apparatus of Pinochet's dictatorship. Following the end of the authoritarian regime in March 1990, Radio Nacional was subject to the new center-left government's media agenda. In 1994, in the middle of Chile's debt crisis, the private Production Development Corporation (CORFO) bought the radio station. Later, Chilean businessman Santiago Agliatti owned it.

9 After its privatization in 1994, Radio Nacional hired a series of media figures linked to Pinochet's civic-military dictatorship, such as Sebastiano Bertoloni, former deputy director of the state newspaper *La Nación*, which was well known as the mouthpiece for the dictatorship's communication agenda. No doubt this new ideological and political milieu of the station aggravated the always latent racial, cultural, and political tensions for the Mapuche radio team.

10 On the feminist community politics of Radio Tierra, see Poblete 2006.

11 By 2013, leading members Ramón Curivil Paillavil and Elías Paillan Coñoepan had already moved to the south, close to or within communities of their Mapuche kinships, near the coastal town of Puerto Saavedra, located east of the city of Temuco, province of Cautín, that is, Lafkenche territory. To some degree, their relocation to the heart of Ngulu Mapu reflected the Mapuche movement's encouragement that the Mapuche diaspora engage in a politics of *retorno*.

12 By this time, José Paillal Huechuqueo and Margarita Elizabeth Huenchual Millaqueo had launched *Wixage Anai* as a space of conversations on Facebook live, where their focus on Mapuche cultural matters constituted virtual echoes of the aims and preoccupations of *Wixage Anai*.

13 The Chilean state promulgated the Indigenous Law on September 28, 1993. In the second paragraph of its Article 1, Title 1, it states, "El Estado reconoce como principales etnias indígenas de Chile a: la Mapuche, Aimara, Rapa Nui o Pascuense, la de las comunidades Atacameñas, Quechuas y Collas del norte del país, las comunidades Kawashkar o Alacalufe y Yámana o Yagán de los canales australes. El Estado valora su existencia por ser parte esencial de las raíces de la nación chilena, así como su integridad y desarrollo, de acuerdo a sus costumbres y valores" (CONADI 1993, 7). In this supposedly "intercultural" statement, the Indigenous Peoples of Chile are hierarchically reduced to "Indigenous ethnicities" (*etnias*) and defined as an "essential part of the roots of the Chilean nation." As such, they are recognized only as the bearers of "ethnic difference" within the framework of the *nación chilena*.

14 In this chapter, when I quote the script or voices from the *Wixage Anai* collective, I use the Raguileo Alphabet to honor their language politics. Nevertheless, for the rest of this chapter—as throughout this book—in my own narrative I mostly use the Mapudungun grammar known as Unified Alphabet.

15 On the corporatization of media in late twentieth-century Chile, see Mönckeberg 2008; and Sunkel and Geoffroy 2001.

16 I purposefully avoid using the term *ritual* and its derivatives since, as a result of the rhetoric of academic indigenismo in Latin American studies and "the anthropology of the Indian" tradition, it has exoticizing resonances when portraying Native religious life. Instead, I use the word *ceremony*, since this term has been much more situated within critical Indigenous studies and scholarship by Native American and Indigenous authors as well as in the language that is used in tribal talk.

17 For further elaboration on the role of music and sound in the Nguillatun ceremony, see Pérez de Arce (2007) 2020, 121–32.

18 Even though the common assumption among many Mapuche at this time was that a Nguillatun (or Gijatun) was difficult to imagine in the urban environment of Santiago, research on Mapuche religious

practices conducted by members of *Wixage Anai* through their Jvfken Mapu Center of Mapuche Communications demonstrated that the organization of this type of ceremony had a long history in the Chilean capital city, and Augusto Aillapan—a Machi—had leading roles in previous Nguillatun that had taken place in Santiago in 1989 and 1990. According to their 2001 study, a Nguillatun took place in the mid-1970s, at the beginning of the Chilean civic-military dictatorship in the O'Higgins Park of Santiago. Under the sponsorship of Radio Colo Colo, this ceremony was organized as a prayer for rain during a critical drought in central Chile. Another Nguillatun had taken place in the Commune of La Florida, in eastern Santiago, on March 18 and 19, 1989. It was organized by the grassroots organization Consejo Mapuche Kiñewküleaiñ Taiñ Rakiduam from the Commune of Cerro Navia. This was the first ceremony of this type convened by Mapuche leaders in the Chilean capital city; among the leading organizers were Misael Alcapan Paillal and Sofía Painequeo in collaboration with José Ancan, María Huichalao, and Machi Augusto Aillapan. This Nguillatun of March 1989 was conceived as a response to the official discourse of the Pinochet-led civic-military dictatorship that aimed to homogenize the population of the country as Chilean only. As one of the leaders of the Consejo Mapuche stated, they wanted to demonstrate that "here we are not all the same and that we are Mapuche and not Chileans." By this time, Pinochet's regime was trying to complete the implementation of its aggressive neoliberal policy of breaking up communal lands into parcels in Mapuche territories. This encountered the resistance of many Mapuche communities. The Nguillatun of March 1989 and subsequent ones in Cerro Navia (in June 1989 and February 1990), in which Machi Aillapan actively participated, were part of the broader resurgent Mapuche movement. See Curivil Paillavil et al. 2001, 48–53.

19 Meli Wixan Mapu was created in Santiago in 1991 as an organization aimed at connecting Mapuche who lived and worked in the Chilean capital city. Kilapan was a collective devoted to supporting communities based in the historic territories of Ngulu Mapu. The Coordinadora Mapuche Región Metropolitana was a coalition that brought together different Mapuche organizations from greater Santiago.

20 As the previously described kull kull and trutruka, the pifilka is also a wind instrument in Mapuche musical culture. The pifilka is a type of flute made of either stone or wood. On this instrument, see Pérez de Arce (2007) 2020, 272–85.

21 Two relevant volumes—Bessire and Fisher, *Radio Fields: Anthropology and Wireless Sound in the 21st Century* (2012), and Wilson and Stewart, *Global Indigenous Media: Cultures, Poetics, and Politics* (2008)—include several articles that focus on the significant role of

radio in Indigenous resurgence and permanence in different regions. However, it is striking to note that, although these volumes were edited by US-based scholars and published by American university presses, neither includes any article on the significant connective role of radio in the Native American experience or within Indigenous Latin American immigrant communities in the United States.

22 On the spread of Mapuche uses of radio, in a previous study I analyze the cases of *Wixage Anai* in Ngulu Mapu and the brief radio shows produced between 2003 and 2005 by the Mapurbe Communication Team in Bariloche, southern Argentina (Puel Mapu). See Cárcamo-Huechante 2013.

23 To learn about the program *First Voices Radio*, see https://www.firstvoicesindigenousradio.org.

24 J. Kēhaulani Kauanui, founder, producer, and host of *Indigenous Politics: From Native New England and Beyond* has published an excellent collection of interviews conducted at her radio show. See Kauanui 2018.

25 On the history of *Contacto Ancestral* as an Indigenous Maya radio show in southern California, see Estrada 2013.

CHAPTER FIVE. ENDURING LISTENING AND SOUNDS

1 In Mexico, *charro* is the name for the "horseman skilled in roping and riding." Charros became known "for their distinctive and flamboyant customs" and "gained fame in the Mexican Revolution because they formed a great part of the insurgent groups," such as those of Pancho Villa and Emiliano Zapata. They also generated the sport known as *charrería*, which combined the practice of *jineta* (Moorish-style riding) "with events derived from cattle ranching" (LeCompte 1996, 83). Regarding the flamboyant style of Mexican charros, in his *Diccionario breve de mexicanismos*, Guido Gómez de Silva describes their typical costumes: "a special suit composed of white shirt, short jacket and a wide brim hat" (2001, 43). In his dictionary entry, Gómez de Silva also refers to the *charras*—Mexican women who became part of this tradition usually dressing in a "wide and long skirt." Since charros originally were part of the administrative and workforce in haciendas—overseers, regular workers—and sometimes owners of ranchos (small or midsize farms), Mexican popular music such as rancheras and corridos came to be associated with charro culture.

2 The Mexican film *Allá en el rancho grande* (1936), which features the corrido of homonymous title, was screened in movie theaters in Chile in 1937, first in Santiago and then in the main cities of the provinces of Ngulu Mapu. There is no doubt that cinema had a role in the dissemination of Mexican popular music in Chile / Ngulu Mapu, but not

at the mass popular scale, across urban and rural areas, as radio did. For a succinct look at the arrival and impact of Mexican cinema in Chile, see Villalobos Dintrans 2012, 86–87. The screening of Mexican movies on television, including rancheras and corridos, would add more to this process, but this medium would attain a real popular boom only by the 1980s. According to Chilean media scholar Valerio Fuenzalida Fernández, studies on media access and consumption in Chile conducted in 1969 showed that among popular and working-class sectors in Chile, radio ranked at the top of the list, followed by newspapers and then revistas (magazines), cinema was barely fourth, television was fifth, and very low in the survey, books barely placed sixth. Furthermore, comparing 1967 and 1987, studies show that by the latter year televisions became present in a high percentage of homes in Chile, especially in cities (Fuenzalida Fernández 1990, 62). In rural areas of southern Chile / Ngulu Mapu, daily access to television started happening to a significant degree only in the late 1990s, and going to movie theaters still was and continued to register today as an urban culture phenomenon. Even though Mapuche migration to cities increased substantially throughout the second half of the twentieth century (Curivil Bravo 2012, 161–66), between the late 1960s and 1980s the incidence of movie theater going among Mapuche urban dwellers remained lower than that of radio and, even, from the 1980s onward, far below television.

3 For Radio SAGO, as a station owned by the Sociedad Agrícola y Ganadera de Osorno (SAGO), "cooperating with the development of the commerce and the industry" in the region has been at the core of its mission (Peralta Vidal and Hipp Troncoso 2004, 215–16).

4 According to the author, who grew up in Rawe Bajo, the old house which functioned as a brothel probably since the mid-twentieth century, still existed in the mid-1970s (Huinao, interview by author, 2024).

5 During the 1960s, the names of Mexican singers and artists such as José Alfredo Jiménez, Pedro Infante, Jorge Negrete, Cuco Sánchez, Antonio Aguilar, Chavela Vargas, Charro Avitia, and Miguel Aceves Mejías, among others, became familiar to increasing number of listeners of charro music among the Mapuche as well as campesino and popular segments of settler Chilean population.

6 As noted in chapter 2, the 1960s as a decade of progressive massification of portable radio receivers throughout Chile has been revealed in monographic media studies of the period (Lasagni et al. 1985, 16).

7 Radio Eleuterio Ramírez was founded in 1958, and Radio Concordia, in 1957. In this period, Radio Pudeto was also launched in Ancud, in 1959; and Radio Camilo Henríquez in Valdivia, in 1958.

8 To problematize stereotypical views on *mexicanidad* and Mexican popular music, which prevailed in scholarships of the 1970s and 1980s

(i.e., Gradante 1982; Geijerstam 1976), Olga Nájera-Ramírez states, "In addition to assuming that indulgence in passion is uniquely Mexican, such readings assume that the ranchera, like the people to whom speaks, is simple, unsophisticated, and therefore transparent in meaning" (2007, 458).

9 According to the canonical view of scholar Américo Paredes, the Mexican corrido "includes the Mexican's spirit of bravado, his exaggerated manliness or 'machismo,' the supreme self-confidence in himself and his own ways" (1963, 233). In recent scholarship, historian Alex Chávez has keenly revealed/untangled the logics of machismo in the performances of *huapango*, another popular musical genre from Mexico, without falling into the trap of essentializing it as part of an allegedly primordial mexicanidad—as it has been naturalized, for example, by thinkers like Octavio Paz. He discusses the machista aspects of the huapango as part of a performative, ideological construction (Chávez 2017, 40–41). In this regard Chávez states, "While certain *canciones típicas* (traditional songs) and *canciones rancheras* (country songs) are overtly melancholic, with expressions of pastoral longing ("Canción Mixteca," for instance), the featured huapango performances most densely embody a near-cartoonlike Mexican aesthetic of machismo, particularly given their placement in presumed masculine spaces: the *palenque* (cockfighting arena) and cantina" (38).

10 María Isabel Anita Carmen de Jesús Vargas Lizano, who later became known as Chavela Vargas, was born in Costa Rica in 1919 and moved to Mexico in the early 1930s. For a study of her persona as a queer, lesbian/butch figure and voice, see Alonso-Minutti 2020.

11 Originally from the central region of Chile, Esmeralda González Letelier emerged as Guadalupe del Carmen on the artistic scene by the early 1950s. As a way to profile herself as a "Chilean" interpreter devoted to cultivating "Mexican" popular music, she adopted the names of two icons of Catholic culture of both countries, that of the widely popular Virgin of Guadalupe from Mexico, and the Lady of Mount Carmel or Virgin of Carmel (Virgen del Carmen) who by the early nineteenth century was officialized as the patron saint of the newly formed Chile nation-state. Through her pseudonym, Guadalupe del Carmen differentiated herself from the musical identity of her husband, Marcial Campos, who was member of the duet Hermanos Campos, influential performers of the Chilean cueca (Creole national dance). Furthermore, toward the first half of the 1980s Guadalupe del Carmen starred in the shows of the Circus of Timoteo, a comic circus founded in Santiago in 1968 mostly comprising cross-dressers and openly queer comedians and popular artists. Participating in this environment of sexual dissidence gave Guadalupe del Carmen a queer inflection to her artistic trajectory.

12 What I underscore about the transformative role of women's listening and singing of Mexican popular music in both the Mapuche and Chilean contexts coincides with scholarship on the reception and performance of the ranchera and corrido genres on both sides of the US-Mexico border. Olga Nájera-Ramírez writes, "As performers, producers and performers of the ranchera, women have increasingly taken a once predominantly male expressive form to give expression to their own desires, needs and experiences" (2007, 474).

13 Radio Eleuterio Ramírez was owned by the Communist Party and Radio Camilo Henríquez by the Socialist Party.

14 In his characterization of what hacienda ownership entails in Mexican society, historian Enrique Semo writes, "La tendencia hacia la autarquía, el dominio del mercado local y la separación del campesino de sus medios de producción, imprimió al hacendado una voracidad inusitada e insaciable de la tierra" (30). For a succinct overview on the formation of the hacienda regime in Mexico, see Semo 2012, 15–35.

15 For a more detailed historical account on the formation of haciendas in Central Chile, see Bauer 1975, 134–41.

16 The Título de Merced, as a land title granted by the Chilean government to relocated or territorially "reduced" Mapuche communities, was the main legal tool to consolidate Mapuche removal within the Chilean settler state's law, especially in the Araucanía region. This legal face of native land dispossession was administered by the Comisión Radicadora de Indígenas (Indigenous Relocation Commission), a state entity created in 1883, starting its work in 1884 and completing it in 1929 (Bengoa 2000, 333–62; Correa Cabrera 2021, 166–70; Mariman Quemenado 2006, 119–22). South to the Araucanía, between the late nineteenth and mid-twentieth centuries, the type of land titles used by state offices to regulate the already-reduced Mapuche-Williche lands ranged from Títulos de Merced to kinship-based titles of "effective possession"—as was the case of my Huechante family in the lands of Tralcao.

17 In his book *Haciendas y campesinos,* José Bengoa offers a detailed account on the establishment of the fundo or *latifundio* in the Araucanía region as well as in the provinces of Valdivia, Osorno, and Llanquihue. See Bengoa 1990, 151–208.

18 In Ngulu Mapu/southern Chile, "chicha" is commonly apple cider with a high alcohol content.

19 On the presence of Mexican popular music in the poetry of Mapuche authors, see López Duhart 2017.

20 The poems I have in mind from Juan Paulo Huirimilla's collection of poetry *Palimpsesto* are "Ranchera de madrugada" (83); "Desvaríos" (86–87); and "Juan de Dios Peñan" (88). As commented, these three poetic texts are traversed by the resonances of Mexican popular music.

21 As previously pointed out, only in the early 2000s Mapuche musicians began to compose ranchera and corrido songs with Mapuche content. Besides the already cited music band Werkenes del Amor, Wechekeche ñi Trawün, the Santiago-based music group we analyze in this section, was also one of the first ones in composing Mapuche ranchera and corrido songs.

22 After the coup d'état of 1973, the Pinochet-led authoritarian regime established curfews as a legal measure to restrict people's circulation in public spaces to discourage any form of "political" gathering. Curfews were common between 1973 and 1975. Until 1987, this such coercive measures were used whenever the regime suspected of oppositional activity in the country.

23 In his doctoral dissertation (2015), ethnomusicologist Jacob Rekedal documents the emergence of rap and hip-hop culture in the Araucania region, and more specifically in Temuco, including a detailed account on the role of the hip-hop group Brocas de las Naquis in this process. See Rekedal 2015, 317–53.

24 Focused on the developments of hip-hop and rapping in Indigenous North America, Kyle Mays's book *Hip Hop Beats, Indigenous Rhymes* (2018) is a pivotal contribution to the study of Indigenous hip-hop.

25 Names in parenthesis are artistic pseudonyms that some of the Wechekeche Ñi Trawün members adopted for their musical performances.

26 I refer here to "Epu Filu Kiñe Füta Weichan," a song included in Wechekeche Ñi Trawün's album titled *Wechekeche Ka Kiñe* (2006).

27 Besides YouTube and Spotify, Wechekeche ñi Trawün have used Facebook and Instagram to disseminate their music and community work.

28 The collective's political and musical work became an important reference point for other Mapuche rappers in the early twenty-first century. Among them, it is worth mentioning Waikil (Jaime Cuyanao) and Luanko (Gonzalo Minuto Soler), who achieved a significant position in the local and transnational hip-hop scene and, at different times, have also been part of Wechekeche Ñi Trawün. Likewise, young Mapuche women rappers—such as MC Millaray, from Santiago; Isleña Antumalen, from Temuco; and Ailla Pumas, from Panguipulli, a town in the northeastern part of the Province of Valdivia—have gained enormous significance in recent years. As with Wechekeche Ñi Trawün, songs by these Mapuche rappers can be found on YouTube.

29 For a scholarly study on Mapuche musical instruments, see Avendaño Curaqueo 2010; Pérez de Arce (2007) 2020.

30 *Ülkantuchefe*: Mapuche person who sings and cultivates the Mapuche genre of ül.

31 Besides the Mapuche participants, Wechekeche Ñi Trawün invited to be part of this collective rap performance a Kaweskar activist, who

participates as a vocalist by near the end of it as a gesture toward the aspiration for a wider alliance of Indigenous voices in what today is known as Chile.

32 Imani Perry writes, "As an art form," rap "combines poetry, prose, song, music and theater" (2004, 38). In attention to the rap performance of Wechekeche ñi Trawün, I certainly point out elements that are more specific to it.

33 Curivil Paillavil suggests the following ül subgenres: *ñiwa-vlkantun* (song of the skillful); *tayvl-vlkantun* (religious song, often practiced by the Machi); *konco-vlkantun* (song of friendship); *laku-vlkantun* (songs that are used in children's naming ceremonies); *gvman-vlkantun* (song of sadness); *gojin-vlkantun* (song of fiestas); *Maci-vlkantun* (chant of the Machi). See Curivil Paillavil 2002, 23. Regarding *tayvl-vlknatun*, or *tayül-ulkantun*, Avendaño Curaqueo states that there are more religious *tayül*, which are the jurisdiction of the Machi; nevertheless, there are also more secular *tayül*, which are practiced by community members to tell their own life stories. She also adds the existence of the *ayekan-ülkantun*, a type of ül that aims to uplift the spirits of people (interview by author, 2023). Regarding recent literature focused on specific case studies, see Jordán and Salazar 2022, whose study recovers from the archive seven songs by ülkantuchefe Juan de Dios Curilem Millanguir, including musical and alphabetic transcriptions as well as reinterpretations of his ül.

34 Félix José de Augusta was born in Augsburg, Germany, in 1860, and died in Valdivia, southern Chile in 1935.

35 It seems that de Augusta's phonographic registers deteriorated or were lost; none of them are available today.

36 "Ngënechen," in Mapudungun, refers to a superior being from the natural universe, a being that even can shape-shift in connection with different aspects and spaces of the Mapu (land, earth, universe). De Augusta debases this concept in his translation by using the Spanish word and concept for God (Dios). To Mapuche scholar Héctor Nahuelpan Moreno, this "homology of the Christian God" with Ngënechen is "part of a process of decontextualization and reduction of the Mapuche religious belief system based on an anthropocentric view that places people as the center of creation and life" (2013, 207). The Mapuche worldview entails a radically different approach, one in which the Mapu is at the center. This is well illustrated in the Mapudungun of Carmen Cumillanka Naquil's chant. For my part, as noted, in my Spanish/English translations, to avoid additional mediating concepts I elect to keep "Ngënechen" in Mapudungun and I proceed on the same terms for pu longko (community chiefs) and for wingka (White settler).

37 Elisa Avendaño Curaqueo was born in 1956 and spent her childhood and adolescence in the Manuel Chavarría Community, in the

countryside near Lautaro, about eighteen and a half miles north of Temuco. The historical process of territorial dispossession in Ngulu Mapu left her family with practically no land. So she emigrated at a very young age to the city of Concepción. In the late 1970s, she returned to Temuco, where she lived up to the early 1990s. In those years, she moved to the outskirts, to the urban area of Padre de Las Casas. Finally, in 2015, she moved to the Francisco Quereban Community, located in the rural sector of Paillanao, in the Commune of Padre de las Casas.

38 Avendaño Curaqueo refers here to the presidential periods of Michelle Bachelet (2006–10; and 2014–18), from the center left wing; Sebastián Piñera (2010–14; and 2018–22), from the center right wing; and General Augusto Pinochet (1973–90), a far right-wing dictatorship (interview by author, 2023).

39 Nowadays, Wiñol Tripantü is also known as We Tripantü. For a deeper analysis of its meaning, see Malo Huencho et al. 2003, 23–35; Curivil Paillavil et al. 2001, 42–43; Pozo Menares and Canio Llanquinao 2015, 108–12.

40 Worth mentioning here is the Rapa Maquehue Festival, a Mapuche music festival which was first held in 2013 in the Maquehue community, near the city of Temuco. This massive, popular festival showcased the musical diversity to which I refer. It took place annually until discontinued during the COVID pandemic.

41 The overlapping and coalescence of Indigenous and non-Indigenous sounds in the musical life of Ngulu Mapu is also reflected in the creative use of Western instruments such as the accordion, bugle, and guitar, not only in Mapuche social gatherings but also, in some communities, in ceremonies. On this topic, see the documentary video *Aukinkoi ñi Vlkantun, hace eco mi música* (Avendaño Curaqueo 2010).

CODA

1 In Ngulu Mapu, last name(s), along with place(s) of origin, are key markers of Mapuche belonging. The former is about our *küpalme* (kinship) and the latter about our tuwün (the specific territory in Wallmapu to which our kinship is historically linked). Therefore, as a Mapuche-Williche, Tralcao and Huechante are critical identifiers of my own family and community genealogy.

2 *Lof* is the basic territorial and communal unit within Mapuche society.

3 The experience of "shame," to which I allude here, iterates as something quite common among Indigenous peoples across different continents. According to Tanana Athabascan scholar Dian Million, shame is a "key component" of what by 1975 Métis academic and activist Howard Adams denominated "internal colonization." With the

experience of "residential school survivors" in Canada as her specific context of reference, Million links the issue of shame to that which she defines as "emotional colonialism," that is, colonialism as "felt, affective relationship" (2013, 46–49).

4 In a conversation years ago, poet and ülkantuchefe Lorenzo Aillapan Cayuleo suggested to me that, for him, Weche Antü would be like the tenuous sun that returns to Ngulu Mapu during the time of Wiñol Tripantu (or Wiñol Xipantu) in late June.

REFERENCES

PUBLICATIONS

Academia Chilena de la Lengua. 1978. *Diccionario del habla chilena*. Santiago: Editorial Universitaria.

Aguirre, Beatriz, and Salim Rabí. 2009. "La trayectoria espacial de la Corporación de la Vivienda (CORVI)." *Diseño Urbano y Paisaje*. Year 6, no. 18: Section 1.4. https://dup.ucentral.cl/pdf/18_trayectoria_espacial_b.pdf.

Aillapan Cayuleo, Lorenzo. 2003. *Üñümche: Hombre pájaro*. Santiago: Pehuén Editores.

Aillapan Cayuleo, Lorenzo. 2017. *Hombre pájaro / Üñümche: Vida y poesía de un Mapuche / Kiñe Mapuche wenttxu ñi mogen ka ñi ulkan*. Santiago: Editora e Imprenta Maval.

Ak'abal, Humberto. [1990] 2008. *El animalero—The animal gathering*. Guatemala City: Editorial Piedra Santa.

Ak'abal, Humberto. 1995. *Hojas del árbol pajarero*. Mexico City: Editorial Praxis.

Albornoz, César. 2014. "La experiencia televisiva en el tiempo de la Unidad Popular: *La Caldera del Diablo*." In *Fiesta y drama: Nuevas historias de la Unidad Popular*, edited by Julio Pinto Vallejos. Santiago: LOM Ediciones.

Alonqueo, Martín. 1985. *Mapuche ayer-hoy*. Padre de las Casas, Temuco: Imprenta y Editorial San Francisco.

Alonso-Minutti, Ana R. 2020. "Chavela's Frida: Decolonial Performativity of the Queer Llorona." In *Decentering the Nation: Music, Mexicanidad, and Globalization*, edited by Jesús A. Ramos-Kittrel. Lanham, MD: Lexington.

Altamirano, Magdalena. 2007. "De la copla al corrido: Influencias líricas en el corrido mexicano tradicional." In *La copla en México*, edited by Aurelio González. Mexico City: El Colegio de México.

Amorós, Celia. 2007. "Thinking Patriarchy." In *Feminist Philosophy in Latin America and Spain*, edited by María Luisa Femenías and Amy A. Oliver. New York: Editions Rodopi B. V.

Anaya, S. James. 1996. *Indigenous Peoples in International Law*. New York: Oxford University Press.

Anderson, Benedict. 1983. *Imagined Communities: Reflections on the Origin and Spread of Nationalism*. London: Verso.

Antileo Baeza, Enrique. 2012. "Migración Mapuche y continuidad colonial." In *Ta iñ fijke xipa rakizuameluwün: Historia, colonialismo y resistencia desde el país Mapuche*. Temuco: Ediciones Comunidad de Historia Mapuche.

Antinao Varas, Clorinda. 2014. *Diccionario ta iñ Mapun Dungun: Nuestra lengua Mapuche*. Santiago: Pu Lifru Mapunche Kimün Ngelu.

Ardito, Lorena, Eileen Karmy, Antonia Mardones, and Alejandra Vargas, eds. 2016. *¡Hagan un trencito! Siguiendo los pasos de la memoria cumbianchera en Chile (1949–1989)*. Santiago: Ceibo Ediciones.

Avendaño Curaqueo, Elisa. 2010. *Aukinkoi ñi vlkantun*. Temuco: Talleres de Imprenta América.

Bauer, Arnold J. 1975. *Chilean Rural Society from the Spanish Conquest to 1930*. Cambridge: Cambridge University Press.

Bengoa, José. 1990. *Hacienda y campesinos: Historia social de la agricultura chilena*. Vol. 2. Santiago: Ediciones Sur.

Bengoa, José. 2000. *Historia del pueblo Mapuche: Siglo XIX y XX*. Santiago: LOM Ediciones.

Bernstein, Charles. 1998. *Close Listening: Poetry and the Performed Word*. New York: Oxford University Press.

Bessire, Lucas, and Daniel Fisher, eds. 2012. *Radio Fields: Anthropology and Wireless Sound in the 21st Century*. New York: New York University Press.

Blest Gana, Alberto. 1860. *La aritmética en el amor: Novela de costumbres*. Valparaíso: Imprenta y Librería del Mercurio de Santos Tornero.

Blest Gana, Alberto. [1862] 1977. *Martín Rivas*. Caracas: Biblioteca Ayacucho.

Blest Gana, Alberto. [1862] 2005. *Mariluán: Un drama en el campo*. Santiago: LOM Ediciones.

Blest Gana, Alberto. 1909. *El loco estero: Recuerdos de la niñez*. Paris: Garnier.

Blest Gana, Alberto. 2011. *Epistolario de Alberto Blest Gana*. Vol. 1. Edited by José Miguel Barros Franco. Santiago: Ediciones de la Dirección de Bibliotecas, Archivos y Museos (DIBAM).

Bourdieu, Pierre. [1998] 2001. *Masculine Domination*. Stanford, CA: Stanford University Press.

Bronfman, Alejandra. 2016. *Isles of Noise: Sonic Media in the Caribbean*. Chapel Hill: University of North Carolina Press.

Buarque de Holanda Ferreira, Aurélio. 1980. *Dicionário da língua portuguesa*. Rio de Janeiro: Editora Nova Frontera.

Calle, Ana Cecilia. 2020. "Al ritmo de la cumbia: Representaciones culturales, literarias y sonoras en Colombia y Argentina." PhD diss., University of Texas at Austin.

Campos, Rubén. 1928. *El folklore y la música mexicana: Investigación acerca de la cultura musical en México (1525–1925)*. Mexico City: Talleres Gráficos de la Nación.

Campt, Tina M. 2017. *Listening to Images*. Durham, NC: Duke University Press.

Cárcamo-Huechante, Luis E. 2006. "Milton Friedman: Knowledge, Public Culture, and Market Economy in the Chile of Pinochet." *Public Culture* 18, no. 2: 413–35.

Cárcamo-Huechante, Luis. 2007. *Tramas del mercado: Imaginación económica, cultura pública y literatura en el Chile de fines del siglo veinte*. Santiago: Editorial Cuarto Propio.

Cárcamo-Huechante, Luis E. 2013. "Indigenous Interference: Mapuche Use of Radio in Times of Acoustic Colonialism." *Latin American Research Review* 48: 50–68.

Cárcamo-Huechante, Luis. 2023. "A Trope of Colonial Obliteration? A Critical Note on 'Colonial Latin America' and Related Conversations." *Colonial Latin American Review* 32, no. 2: 243–48.

Cárcamo-Huechante, Luis, and Elías Paillan Coñoepan. 2012. "Taiñ pu amulzugue egvn: Sonidos y voces del Wajmapu en el aire." In *Ta iñ fijke xipa rakizuameluwün: Historia, colonialismo y resistencia desde el país Mapuche*. Temuco: Ediciones Comunidad de Historia Mapuche.

Carrasco Muñoz, Hugo. 2002. "Rasgos identitarios de la poesía Mapuche actual." *Revista Chilena de Literatura*, no. 61: 83–110.

Castillo, Debra A., and Shalini Puri. 2016. "Introduction: Conjectures on Undisciplined Research." In *Theorizing Fieldwork in the Humanities: Methods, Reflections, and Approaches to the Global South*, edited by Shalini Puri and Debra A. Castillo. New York: Palgrave Macmillan.

Catrileo, María. 2010. *La lengua Mapuche en el siglo XXI*. Valdivia: Facultad de Filosofía y Humanidades, Universidad Austral de Chile.

Catrileo, María. 2017. *Diccionario lingüístico etnográfico de la lengua Mapuche*. Valdivia: Ediciones UACH.

Charny, Israel W. 1994. "Toward a Generic Definition of Genocide." In *Genocide: Conceptual and Historical Dimensions*, edited by George J. Andreopoulos. Philadelphia: University of Pennsylvania Press.

Chávez, Alex. 2017. *Sounds of Crossing: Music, Migration, and the Aural Poetics of Huapango Arribeño*. Durham, NC: Duke University Press.

Chion, Michel. [1990] 1994. *Audio-Vision: Sound on Screen*. Edited and translated by Claudia Gorbman. New York: Columbia University Press.

Colectivo Editorial Mapuexpress. 2016. *Resistencias Mapuche al extractivismo*. Santiago: Editorial Quimantú.

Collier, Simon, and William Sater. 1996. *A History of Chile, 1808–1994*. Cambridge: Cambridge University Press.

Comunidad de Historia Mapuche. 2012a. "Introduction." In *Ta iñ fijke xipa rakizuameluwün: Historia, colonialismo y resistencia desde el país Mapuche*. Temuco: Ediciones Comunidad de Historia Mapuche.

Comunidad de Historia Mapuche. 2012b. *Ta iñ fijke xipa rakizuameluwün: Historia, colonialismo y resistencia desde el país Mapuche*. Temuco: Ediciones Comunidad de Historia Mapuche.

Comunidad de Historia Mapuche. 2015. *Awükan ka kuxankan zugu Wajmapu mew: Violencias coloniales en Wajmapu*. Temuco: Ediciones Comunidad de Historia Mapuche.

CONADI (Corporación Nacional de Desarrollo Indígena). 1993. *Ley Indígena: Ley N° 19.253*. Santiago: Ministerio de Planificación y Cooperación, Gobierno de Chile.

Concha, Jaime. 2011. *Leer a contraluz: Estudios sobre narrativa chilena. De Blest Gana a Varas y Bolaño*. Santiago: Ediciones Universidad Alberto Hurtado.

Correa Cabrera, Martín. 2021. *La historia del despojo: El origen de la propiedad particular en el territorio Mapuche*. Santiago: Pehuén Editores / Ceibo Ediciones.

Coulthard, Glen Sean. 2014. *Red Skin, White Masks: Rejecting the Colonial Politics of Recognition*. Minneapolis: University of Minnesota Press.

Crow, Joanna. 2013. *The Mapuche in Modern Chile: A Cultural History*. Gainesville: University Press of Florida.

Cuddon, J. A. 1998. *The Penguin Dictionary of Literary Terms and Literary Theory*. 4th ed. New York: Penguin.

Curivil Bravo, Felipe Domingo. 2012. "Asociatividad Mapuche en el espacio urbano Santiago, 1940–1970." In *Ta iñ fijke xipa rakizuameluwün: Historia, colonialismo y resistencia desde el país Mapuche*. Temuco: Ediciones Comunidad de Historia Mapuche.

Curivil Paillavil, Ramón Francisco. 2002. *Lenguaje Mapuce para la educación intercultural: Aproximaciones al Mapucezugun*. Santiago: Talleres de Gráfica Andes.

Curivil Paillavil, Ramón Francisco, Carmen Malo Huencho, José Alfredo Paillal Huechuqueo, and Elías Paillán Coñoepan. 2001. *Identidad Mapuche y prácticas religiosas tradicionales en Santiago*. Santiago: Centro de Comunicaciones Jvfken Mapu.

de Augusta, Félix. 1934, *Lecturas Araucanas*. 2nd ed. Padre de las Casas, Province of Cautín: Imprenta y Editorial "San Francisco."

Dennison, Philip J. 1998. *Playing Indian*. New Haven, CT: Yale University Press.

Del Solar, Alberto. 1888. *Huincahual*. Paris: Pedro Roselli Editor.

Dennison, Jean. 2012. *Colonial Entanglement: Constituting a Twenty-First-Century Osage Nation*. Chapel Hill: University of North Carolina Press.

Denning, Michael. 2015. *Noise Uprising: The Audiopolitics of a World Musical Revolution*. New York: Verso.

Derince, Mehmet Şerif. 2013. "A Break or Continuity? Turkey's Politics of Kurdish Language in the New Millennium." *Dialectical Anthropology* 37, no. 1: 145–52.

Dolar, Mladen. 2006. *A Voice and Nothing More*. Cambridge, MA: MIT Press.

Ehrick, Christine. 2015. *Radio and the Gendered Soundscape: Women and Broadcasting in Argentina and Uruguay, 1930–1950*. Cambridge: Cambridge University Press.

Eidsheim, Nina Sun. 2015. *Sensing Sound: Singing and Listening as Vibrational Practice*. Durham, NC: Duke University Press.

Encina A., Francisco. 1954. *Resumen de la historia de Chile*. Vol. 2, 2nd ed. Santiago: Editorial Zig Zag.

Ercilla y Zúñiga, Alonso de. 1993. *La Araucana*. Madrid: Ediciones Cátedra.

Ercilla y Zúñiga, Alonso de. 2004. *Para gozar La Araucana de Alonso de Ercilla y Zúñiga Precursor de Chile*. Selection and study by Herman Schwember and Adriana Azócar. Santiago: Ediciones Tierra Mía / Biblioteca del Bicentenario.

Estévez Trujillo, Mayra. 2008. *Estudios sonoros: Desde la región andina / Sonic Studies: From the Andean Region*. Quito: Centro Experimental Oído Salvaje.

Estrada, Alicia Ivonne. 2013. "Ka Tzij: The Maya Diasporic Voices from Contacto Ancestral." *Latino Studies* 11, no. 2: 208–27.

Fanon, Frantz. [1959] 1967. *A Dying Colonialism*. Translated by Haakon Chevalier. New York: Grove. Originally published as *L'an V de la Révolution Algérienne*. Paris: Cahiers Libres.

Feld, Steven. 1982. *Sound and Sentiment: Birds, Weeping, Poetics, and Song in Kaluli Expression*. Philadelphia: University of Pennsylvania Press.

Fernando, S. H., Jr. 1999. "Back in the Day, 1975–1979." In *The Vibe History of Hip Hop*, edited by Alan Light. New York: Random House.

Fiol-Matta, Licia. 2017. *The Great Woman Singer: Gender and Voice in San Juan Puerto Rican Music*. Durham, NC: Duke University Press.

Fuenzalida Fernández, Valerio. 1990. *La televisión en los '90*. Santiago: Corporación de Promoción Universitaria.

Furlonge, Nicole Brittingham. 2018. *Race Sounds: The Art of Listening in African American Literature*. Iowa City: University of Iowa Press.

Gamboa, Luis. 2000. "Los radioteatros." In *La historia de la radiotelefonía*, broadcasting by Luis Gamboa, Marcelo Jiménez Sufan, and Gina Zuanic. Audiobook (MP3).

García Barrera, Mabel. 2008. "Entre-textos: La dimensión dialógica e intercultural del discurso poético Mapuche." *Revista Chilena de Literatura*, no. 72: 29–70.

Gaudio, Michael. 2019. *Sound, Image, Silence: Art and the Aural Imagination in the Atlantic World*. Minneapolis: University of Minnesota Press.

Geijerstam, Claes af. 1976. *Popular Music in Mexico*. Albuquerque: University of New Mexico Press.

Gómez-Barris, Macarena. 2017. *The Extractive Zone: Social Ecologies and Decolonial Perspectives*. Durham, NC: Duke University Press.

Gómez de Silva, Guido. 2001. *Diccionario breve de mexicanismos*. Mexico City: Academia Mexicana, Fondo de Cultura Económica.

Góngora Marmolejo, Alonso de. 1862. *Historia de Chile: Desde su descubrimiento hasta el año de 1575*. Santiago: Imprenta del Ferrocarril.

Gotschlich, Guillermo. 1991. "Alberto Blest Gana y su novela histórica." *Revista Chilena de Literatura*, no. 38: 29–58.

Gradante, William. 1982. "'El Hijo del Pueblo': José Alfredo Jiménez and the Mexican 'Canción Ranchera.'" *Latin American Music Review* 3, no. 1: 36–59.

Guerra, Lucía. 2014. *La ciudad ajena: Subjetividades de origen Mapuche en el espacio urbano*. Santiago: Ceibo Ediciones.

Haughney, Diane. 2006. *Neoliberal Economics, Democratic Transition, and Mapuche Demands for Rights in Chile*. Gainesville: University Press of Florida.

Henríquez, Camilo. 1812. "Prospecto." In *Aurora de Chile: Periódico Ministerial, y Político*. Santiago: Gobierno de Chile.

Hernández, José. [1872; 1879] 2001. *Martín Fierro*. Critical edition by Elida Lois and Angel Núñez. Nanterre, France: Colección Archivos, Université Paris X.

Higa, Ben. 1999. "Early Los Angeles Hip Hop." In *The Vibe History of Hip Hop*, edited by Alan Light. New York: Random House.

Hofflinger, Alvaro, Héctor Nahuelpan, Alex Boso, and Pablo Millalen. 2021. "Do Large-Scale Forestry Companies Generate Prosperity in Indigenous Communities? The Socioeconomic Impacts of Tree Plantations in Southern Chile." *Human Ecology*, no. 49: 619–30.

Hosiasson, Laura Janina. 2017. "Siete novelas de Blest Gana: Una visión de conjunto." *Revista Chilena de Literatura*, no. 95: 235–58.

Huenún, Jaime Luis. 1999. *Ceremonias*. Santiago: Universidad de Santiago.

Huidobro, Vicente. 1931. *Altazor*. Madrid: Compañía Ibero Americana de Publicaciones.

Huinao, Graciela. 2010. *Desde el fogón de una casa de putas Williche*. Osorno, Chile: Ediciones Caballo de Mar—CONADI.

Huinca Piutrin, Herson. 2012. "Los Mapuche del Jardín de Aclimatación de París en 1883: Objetos de la ciencia colonial y políticas de investigación contemporánea." In *Ta iñ fijke xipa rakizuameluwün: Historia, colonialismo y resistencia desde el país Mapuche*. Temuco: Ediciones Comunidad de Historia Mapuche.

Huirimilla, Juan Paulo. 2005. *Palimpsesto*. Santiago: LOM Ediciones.

Hu Pegues, Juliana. 2021. *Space-Time Colonialism: Alaska's Indigenous and Asian Entanglements*. Chapel Hill: University of North Carolina Press.

Jofré, Manuel Alcides. 1983. *La historieta en Chile en la última década*. Santiago: CENECA.

Jordán, Laura, and Andrea Salazar. 2022. *Trafülkantun: Cantos cruzados entre Garrido y Curilem*. Santiago: Ariadna Ediciones.

Kaempfer, Alvaro. 2006. "Alencar, Blest Gana y Galván: Narrativas de Exterminio y Subalternidad." *Revista Chilena de Literatura*, no. 69: 89–106.

Kauanui, J. Kēhaulani. 2016. "'A Structure, Not an Event': Settler Colonialism and Enduring Indigeneity." *Lateral* 5, no. 1. https://csalateral.org/issue/5-1/forum-alt-humanities-settler-colonialism-enduring-indigeneity-kauanui.

Kauanui, J. Kēhaulani. 2018. *Speaking of Indigenous Politics: Conversations with Activists, Scholars and Tribal Leaders*. Minneapolis: University of Minnesota Press.

Keme, Emil'. 2018. "For Abiayala to Live, the Americas Must Die: Toward a Transhemispheric Indigeneity." *Native American and Indigenous Studies* 5, no. 1: 42–68.

Kheshti, Roshanak. 2015. *Modernity's Ear: Listening to Race and Gender in World Music*. New York: New York University Press.

Klein, Naomi. 2007. *The Shock Doctrine: The Rise of Disaster Capitalism*. New York: Metropolitan / Henry Holt.

Klubock, Thomas Miller. 2014. *La Frontera: Forests and Ecological Conflict in Chile's Frontier Territory*. Durham, NC: Duke University Press.

Kun, Josh. 2005. *Audiotopia: Music, Race, and America*. Berkeley: University of California Press.

LaBelle, Brandon. 2010. *Acoustic Territories: Sound Culture and Everyday Life*. London: Continuum International.

LaBelle, Brandon. 2018. *Sonic Agency: Sound and Emergent Forms of Resistance*. London: Goldsmiths.

Laime Ajacopa, Teofilo. 2007. *Diccionario bilingüe: Iskay simipi yuyayk'ancha*. 2nd ed. La Paz, Bolivia. Self-published. https://futatraw.ourproject.org/descargas/DicQuechuaBolivia.pdf.

Lara, Horacio. 1889. *Crónica de la Araucania: Descubrimiento i conquista, pacificacion definitiva i campaña de Villa-Rica (leyenda heroica de tres siglos)*. Santiago: Imprenta El Progreso.

Lasagni, María Cristina, Paula Edwards, and Josiane Bonnefoy. 1985. *La radio en Chile (historias, modelos, perspectivas)*. Santiago: CENECA.

Láscar, Amado J. 2003. "*Mariluán* y el problema de la inserción del mundo indígena al Estado nacional: Expansión del Estado Nación y rearticulación simbólica del cuerpo indígena." *Working Paper Series*, no. 16. Ñuke Mapuförlaget.

Latcham, Ricardo. 1959(?). *Blest Gana y la novela realista*. Santiago: Ediciones de los Anales de la Universidad de Chile.

LeBonniec, Fabien. 2006. "Rebeliones en La Araucanía: Vueltas y revueltas de la historia a la orilla del Lago Budi." *Anales de Desclasificación* 1, no. 2: 553–65.

LeCompte, Mary Lou. 1996. "Charro." In *Encyclopedia of Latin American History and Culture*, vol. 2, edited by Barbara A. Tenenbaum. New York: Simon and Schuster Macmillan.

Levil Chicahual, Rodrigo. 2006. "La sociedad Mapuche contemporánea . . ." In *¡. . . Escucha, winka . . . ! Cuatro ensayos de historia nacional Mapuche y un epílogo sobre el futuro*, edited by Pablo Mariman, Sergio Caniuqueo, José Millalen, and Rodrigo Levil. Santiago: LOM Ediciones.

Leyva Solano, Xochitl, and Shannon Speed. 2008. "Hacia la investigación descolonizada: Nuestra experiencia de co-labor." In *Gobernar (en) la diversidad: Experiencias indígenas desde América Latina. Hacia la investigación de co-labor*, edited by Xochitl Leyva, Araceli Burguete, and Shannon Speed. Mexico City: CIESAS/FLACSO.

Lienlaf, Leonel. 1989. *Se ha despertado el ave de mi corazón*. Santiago: Editorial Universitaria.

Lienlaf, Leonel. 2003. *Pewma dungu / Palabras soñadas*. Santiago: LOM Ediciones.

López Duhart, María Ignacia. 2017. "Yo vengo a cantar un corrido: El salteador de caminos en la poesía Mapuche Williche." *Revista Chilena de Literatura*, no. 95: 85–105.

Ludmer, Josefina. [1988] 2000. *El género gauchesco: Un tratado sobre la patria*. Buenos Aires: Libros Perfil.

Mackenzie, Adrian. 2002. *Transductions: Bodies and Machines at Speed*. London: Continuum International.

Malo Huencho, Carmen, Margarita Elizabeth Huenchual, José Alfredo Paillal Huechuqueo, and Ramón Francisco Curivil Paillavil. 2003. *Celebración del "Wiñol xipantu": Inicio de un nuevo ciclo de la naturaleza*. Santiago: Centro de Comunicaciones Mapuce Jvfken Mapu.

Mariman Quemenado, Pablo. 2006. "Los Mapuche antes de la conquista militar chileno argentina." In *¡. . . Escucha, winka . . . ! Cuatro ensayos de historia nacional Mapuche y un epílogo sobre el futuro*, edited by Pablo Mariman, Sergio Caniuqueo, José Millalen, and Rodrigo Levil. Santiago: LOM Ediciones.

Mariman Quemenado, Pablo. 2023. "Aspectos fundantes de las relaciones contemporáneas de Wallmapu, Chile y Argentina." *Estudios Sociales* 64, no. 1, year 33 (January–June). https://bibliotecavirtual.unl.edu.ar/publicaciones/index.php/EstudiosSociales/article/view/12856/17641.

Marsilli-Vargas, Xochitl. 2022. *Genres of Listening: An Ethnography of Psychoanalysis in Buenos Aires*. Durham, NC: Duke University Press.

Martín-Barbero, Jesús. [1987] 1993. *Communication, Culture and Hegemony: From the Media to Mediations*. Translated by Elizabeth Fox and Robert A. White. London: SAGE.

Martínez, Juan Luis. 1977. *La nueva novela*. Santiago: Ediciones Archivo.

Masiello, Francine. 1992. *Between Civilization and Barbarism: Women, Nation, and Literary Culture in Modern Argentina.* Lincoln: University of Nebraska Press.

Masiello, Francine. 2013. *El cuerpo de la voz (poesía, ética y cultura).* Rosario: Beatriz Viterbo Editora.

Mays, Kyle. 2018. *Hip Hop Beats, Indigenous Rhymes: Modernity and Hip Hop in Indigenous North America.* Albany: State University of New York Press.

Méndez Montero, Verónica, Carolina Santelices Ariztía, Rodrigo Martínez Iturriaga, and Isidora Puga Serrano. 2009. *Historia, geografía y ciencias sociales. Texto para el estudiante.* Santiago: Ediciones Santillana.

Mendoza, Vicente T. 1964. *Lírica narrativa de México: El corrido.* Mexico City: Instituto de Investigaciones Estéticas; Universidad Nacional Autónoma de México.

Mendoza, Vicente T. [1954] 1976. *El corrido mexicano.* Mexico City: Fondo de Cultura Económica.

Meneses, Lalo. 2014. *Reyes de la jungla: Historia visual de Panteras Negras.* Santiago: Ocho Libros Editores.

Merayo Pérez, Arturo. 2007. "La estimulante diversidad de la radio iberoamericana." In *La radio en Iberoamérica: Evolución, diagnóstico y prospectiva,* edited by Arturo Merayo Pérez. Seville: Comunicación Social Ediciones y Publicaciones.

Merino, María E., and Daniel Quilaqueo. 2003. "Ethnic Prejudice Against the Mapuche in Chilean Society as a Reflection of the Racist Ideology of the Spanish Conquistadors." *American Indian Culture and Research Journal* 27, no. 4: 105–16.

Millalen Paillal, José. 2006. "La sociedad Mapuche prehispánica: *Kimün,* arqueología y etnohistoria." In *¡ . . . Escucha, winka . . . ! Cuatro ensayos de historia nacional Mapuche y un epílogo sobre el futuro,* edited by Pablo Mariman, Sergio Caniuqueo, José Millalen, and Rodrigo Levil. Santiago: LOM Ediciones.

Millán, Moira. 2024. *Terricidio: Sabiduría ancestral para un mundo alternativo.* Buenos Aires: Sudamericana.

Million, Dian. 2013. *Therapeutic Nations: Healing in an Age of Indigenous Human Rights.* Tucson: University of Arizona Press.

Mönckeberg, María Olivia. 2008. *Los magnates de la prensa: Concentración de los medios de comunicación en Chile.* Santiago: Random House Mondadori.

Mora, Mariana. 2017. *Kuxlejal Politics: Indigenous Autonomy, Race, and Decolonizing Research in Zapatista Communities.* Austin: University of Texas Press.

Morales Pettorino, Félix, and Oscar Quiroz Mejías. 1987. *Diccionario ejemplificado de chilenismos y de otros usos diferenciales del español de*

Chile. Vol. 4. Valparaíso: Universidad de Playa Ancha de Ciencias de la Educación.

Morgan, Marcyliena. 2009. *The Real Hiphop: Battling for Knowledge, Power, and Respect in the LA Underground*. Durham, NC: Duke University Press.

Moulián, Tomás. 1997. *Chile actual: Anatomía de un mito*. Santiago: LOM Ediciones.

Mudimbe, V. Y. 1994. *The Idea of Africa*. Bloomington: Indiana University Press.

Murillo, Mario. 2008. "Weaving a Communication Quilt in Colombia: Civil Conflict, Indigenous Resistance, and Community Radio in Northern Cauca." In *Global Indigenous Media: Cultures, Poetics, and Politics*, edited by Pamela Wilson and Michelle Stewart. Durham, NC: Duke University Press.

Nahuelpan, Héctor. 2016. "The Place of the 'Indio' in Social Research: Considerations from Mapuche History." *AlterNative* 12, no. 1: 3–17.

Nahuelpan Moreno, Héctor Javier. 2013. "Wingkün ka kisugünewün: Colonialismo, despojo y agencias históricas Mapuche en NguluMapu." PhD diss., Centro de Investigaciones y Estudios Superiores en Antropología Social (CIESAS), Mexico City.

Nájera-Ramírez, Olga. 2007. "Unruly Passions: Poetics, Performance, and Gender in the Ranchera Song." In *Women and Migration in the US-Mexico Borderlands: A Reader*, edited by Denise A. Segura and Patricia Zavella. Durham, NC: Duke University Press.

Nancy, Jean-Luc. [2002] 2007. *Listening*. New York: Fordham University Press.

Naranjo Villegas, Abel. 1965. "Chilenismos de uso corriente." *Thesaurus* 20, no. 3: 607–11.

Navarro Cofré, Guillermo. 2023. *Beat-Box: Historia del hip-hop en Chile 1984–1994*. Santiago: Autoedition.

Negrete, Jorge. 2002. "Juan Charrasqueado" [Scarface Juan]. In *Corridos sin fronteras: Cancionero / Ballads Without Borders: Songbook*, lyric transcriptions, translations, and notes by Guillermo E. Hernández et al. Washington, DC: Smithsonian Institution.

Neruda, Pablo. 1966. *Arte de pájaros*. Santiago: Ediciones Sociedad de Amigos del Arte Contemporáneo.

Neruda, Pablo. 1971. "El mensajero." In *Don Alonso de Ercilla Inventor de Chile*. Santiago: Editorial Pomaire.

Ochoa, Juan Sebastián. 2016. "La cumbia en Colombia: Invención de una tradición." *Revista Musical Chilena* 70, no. 226: 31–52.

Ochoa Gautier, Ana María. 2014. *Aurality: Listening and Knowledge in Nineteenth-Century Colombia*. Durham, NC: Duke University Press.

Oxford University Press. 2007. *Shorter Oxford English Dictionary: On Historical Principles*. Vol. 1, 6th ed. New York: Oxford University Press.

Pairican Padilla, Fernando. 2015. "El retorno de un viejo actor político: El guerrero. Perspectivas para comprender la violencia política en el movimiento Mapuche (1990–2010)." In *Awükan ka kuxankan zugu WajMapu mew: Violencias coloniales en WajMapu*, edited by Comunidad de Historia Mapuche. Temuco: Ediciones Comunidad de Historia Mapuche.

Paredes, Américo. 1963. "The Ancestry of Mexico's Corridos: A Matter of Definitions." *Journal of American Folklore* 76, no. 301: 231–35.

Pastene, Margarita. 2007. "La radio en Chile." In *La radio en Iberoamérica: Evolución, diagnóstico y prospectiva*, edited by Arturo Merayo Pérez. Seville: Comunicación Social Ediciones y Publicaciones.

Peralta Vidal, Gabriel, and Roswitha Hipp Troncoso. 2004. *Historia de Osorno: Desde los inicios del poblamiento hasta la transformación urbana del siglo XX*. Osorno: Municipalidad de Osorno.

Perea, Jessica Bissett. 2021. *Sound Relations: Native Ways of Doing Music History in Alaska*. New York: Oxford University Press.

Pérez de Arce, José. [2007] 2020. *Música Mapuche*. Santiago: Ocho Libros.

Pérez Galdós, Benito. [1887] 1986. *Fortunata y Jacinta*. Madrid: Taurus.

Perloff, Marjorie. 2009. "The Sound of Poetry." In *The Sound of Poetry / The Poetry of Sound*, edited by Marjorie Perloff and Craig Dworkin. Chicago: University of Chicago Press.

Perloff, Marjorie, and Craig Dworkin, eds. 2009. *The Sound of Poetry / The Poetry of Sound*. Chicago: University of Chicago Press.

Perry, Imani. 2004. *Prophets of the Hood: Politics and Poetics in Hip Hop*. Durham, NC: Duke University Press.

Pinto Rodríguez, Jorge. 2003. *La formación del estado y la nación, y el pueblo Mapuche: De la inclusión a la exclusión*. Santiago: Dirección de Bibliotecas, Archivos y Museos, Centro de Investigaciones Diego Barros Arana.

Poblete, Juan. 2006. "Culture, Neo-liberalism and Citizen Communication: The Case of Radio Tierra in Chile." *Global Media and Communication* 2, no. 3: 315–33.

Poma de Ayala, Felipe Guaman. [1615] 1980. *Nueva corónica y buen gobierno*. Vol. 1. Transcription, prologue, and chronology by Franklin Pease. Caracas: Biblioteca Ayacucho.

Poma de Ayala, Felipe Guaman. 1978. *Letter to a King: A Peruvian Chief's Account of Life Under the Incas and Under Spanish Rule*. Edited and translated by Christopher Dilke. New York: E. P. Dutton.

Pozo Menares, Gabriel, and Margarita Canio Llanquinao. 2015. *Wenumapu: Astronomía y cosmología mapuche*. Santiago: Ocho Libros Editores.

Preminger, Alex, et al. 1993. "Performance." In *The New Princeton Encyclopedia of Poetry and Poetics*. Princeton, NJ: Princeton University Press, 1993.

Raheja, Michelle. 2010. *Reservation Reelism: Redfacing, Visual Sovereignty, and Representations of Native Americans in Film*. Lincoln: University of Nebraska Press.

Real Academia de la Lengua Española. 2001. *Diccionario de la lengua española*. Vol. 2, 22nd ed. Madrid: Editorial Espasa Calpe.

Reed, Anthony. 2021. *Soundworks: Race, Sound, and Poetry in Production*. Durham, NC: Duke University Press.

Reed, Trevor G. 2019. "Sonic Sovereignty: Performing Hopi Authority in Öngtupqa." *Journal of the Society for American Music* 3, no. 4: 508–30.

Rekedal, Jacob Eric. 2015. "Warrior Spirit: From Invasion to Fusion Music in the Mapuche Territory of Southern Chile." PhD diss., University of California, Riverside.

Reschio, Antonio de. 2018. *La Araucanía: Memorias inéditas de la misión capuchina en Chile.* Santiago: Ofqui Editores.

Richards, Patricia. 2013. *Race and the Chilean Miracle: Neoliberalism, Democracy, and Indigenous Rights.* Pittsburgh: University of Pittsburgh Press.

Rivera Cusicanqui, Silvia. 2010. *Ch'ixinakax Utxiwa: Una reflexión sobre prácticas y discursos colonizadores.* Buenos Aires: Tinta Limón.

Robinson, Dylan. 2020. *Hungry Listening: Resonant Theory for Indigenous Sound Studies.* Minneapolis: University of Minnesota Press.

Rojas, Waldo. 1997. *La Araucana de Alonso de Ercilla y Zúñiga y la fundación legendaria de Chile.* Paris: Diffusion Les Belles Lettres.

Rojas Flores, Jorge. 2016. *Las historietas en Chile, 1962–1982: Industria, ideología y prácticas sociales.* Santiago: LOM Ediciones.

Román, Manuel Antonio. 1913–16. *Diccionario de chilenismos y de otras voces y locuciones viciosas.* Vol. 4. Santiago: Imprenta de San José.

Rose, Tricia. 1994. *Black Noise: Rap Music and Black Culture in Contemporary America.* Hanover, NH: Wesleyan University Press.

Rossel, César Enrique. 1965a. *Residencial La Pichanga.* Year 1, no. 3.

Rossel, César Enrique. 1965b. *Residencial La Pichanga.* Year 1, no. 11.

Schafer, R. Murray. [1977] 1994. *The Soundscape: Our Sonic Environment and the Tuning of the World.* Rochester, VT: Destiny Books.

Semo, Enrique. 2012. "Introducción." In *Siete ensayos sobre la hacienda mexicana, 1780–1880*, edited by Enrique Semo. Mexico City: Universidad Nacional Autónoma de México; Instituto Nacional de Antropología e Historia.

Sepúlveda, Alfredo. 2018. *Breve historia de Chile: De la última glaciación a la última revolución.* Santiago: Penguin Random House.

Shoemaker, Nancy. 2015. "A Typology of Colonialism." *Perspectives on History* 53, no. 7. https://www.historians.org/perspectives-article/a-typology-of-colonialism-october-2015.

Silva Castro, Raúl. 1955. *Alberto Blest Gana 1830–1920.* Santiago: Empresa Editora Zig Zag.

Smith, Linda Tuhiwai. [1999] 2008. *Decolonizing Methodologies: Research and Indigenous Peoples.* London: Zed.

Sommer, Doris. 1991. *Foundational Fictions: The National Romances of Latin America.* Berkeley: University of California Press.

Speed, Shannon. 2017. "Structures of Settler Capitalism in Abya Yala." *American Quarterly* 69, no. 4: 783–90.

Steiner, George. 1967. *Language and Silence: Essays on Language, Literature and the Inhuman.* New York: Atheneum.

Sterne, Jonathan. 2003. *The Audible Past: Cultural Origins of Sound Reproduction.* Durham, NC: Duke University Press.
Stewart, Susan. 2002. *Poetry and the Fate of the Senses.* Chicago: University of Chicago Press.
Stoever, Jennifer Lynn. 2016. *The Sonic Color Line: Race and the Cultural Politics of Listening.* New York: New York University Press.
Strong, Pauline. 2004. "The Mascot Slot: Cultural Citizenship, Political Correctness, and Pseudo-Indian Sports Symbols." *Journal of Sports and Social Issues* 28, no. 1: 79–87.
Subercaseaux, Bernardo. 2021. "La Araucana: Un texto que genera contexto." *Nueva Revista del Pacífico*, no. 74: 143–69. http://www.nuevarevistadelpacifico.cl/index.php/NRP/article/view/205.
Sunkel, Guillermo, and Esteban Geoffroy. 2001. *Concentración económica de los medios de comunicación.* Santiago: LOM Ediciones.
Szendy, Peter. 2007. *Sur écoute: Esthétique de l'espionnage.* Paris: Les Éditions de Minuit.
Tahmahkera, Dustin. 2008. "Custer's Last Sitcom: Decolonized Viewing of the Sitcom's 'Indian.'" *American Indian Quarterly* 32, no. 3: 324–51.
Taylor, Diana. 2016. *Performance.* Durham, NC: Duke University Press.
Thompson, Ayanna. 2021. *Blackface.* New York: Bloomsbury Academic.
Toribio Medina, José. 1928. *Chilenismos.* Santiago: Sociedad Imprenta y Litografía Universo.
Turin, Mark. 2012. "Voices of Vanishing Worlds: Endangered Languages, Orality, and Cognition." *Análise Social* 47, no. 205: 846–69.
Valdés, Juan Gabriel. 1995. *Pinochet's Economists: The Chicago School of Economics in Chile.* Cambridge: Cambridge University Press.
Valdivia, Pedro de. [1545–52] 1970. *Cartas de relación de la conquista de Chile.* Santiago: Editorial Universitaria.
Vaziri, Parisa. 2021. "Thaumaturgic, Cartoon Blackface." *Lateral* 10, no. 1. https://csalateral.org/forum/cultural-constructions-race-racism-middle-east-north-africa-southwest-asia-mena-swana/thaumaturgic-cartoon-blackface-vaziri.
Vazquez, Alexandra T. 2013. *Listening in Detail: Performances of Cuban Music.* Durham, NC: Duke University Press.
Vega, Inca Garcilaso de la. 1966. *Royal Commentaries of the Incas and General History of Peru. Part One.* Austin: University of Texas Press.
Veracini, Lorenzo. 2010. *Settler Colonialism: A Theoretical Overview.* London: Palgrave Macmillan.
Vergara-Perucich, Francisco, and Camillo Boano. 2021. "The Big Band of Neoliberal Urbanism: The Gigantomachy of Santiago's Urban Development." *Politics and Space* 39, no. 1: 184–203.
Villalobos, Sergio R. 1983. *Breve historia de Chile.* Santiago: Editorial Universitaria.

Villalobos, Sergio R. 1996. *A Short History of Chile*. Santiago: Editorial Universitaria.

Villalobos Dintrans, Luis. 2012. "Los filmes musicales extranjeros y su influencia en la creación de identidades en el cine y en los espectadores chilenos 1930–1950." *Revista Faro* 1, no. 16: 81–91.

Viñas, David. [1982] 2003. *Indios, ejército y frontera*. Buenos Aires: Santiago Arcos Editor.

Vitale, Luis. 2011. *Interpretación marxista de la historia de Chile*. Vol. 2. Santiago: LOM Ediciones.

Voegelin, Salomé. 2010. *Listening to Noise and Silence: Towards a Philosophy of Sound Art*. New York: Continuum.

Webster, Anthony. 2009. *Explorations in Navajo Poetry and Poetics*. Albuquerque: University of New Mexico Press.

Weheliye, Alexander G. 2014. *Habeas Viscus: Racializing Assemblages, Biopolitics, and Black Feminist Theories of the Human*. Durham, NC: Duke University Press.

Wilson, Pamela, and Michelle Stewart, eds. 2008. *Global Indigenous Media: Cultures, Poetics, and Politics*. Durham, NC: Duke University Press.

Wolfe, Patrick. 1999. *Settler Colonialism and the Transformation of Anthropology: The Politics and Poetics of an Ethnographic Event*. London: Cassell.

Wolfe, Patrick. 2006. "Settler Colonialism and the Elimination of the Native." *Journal of Genocide Research* 8, no. 4: 387–409.

Zúñiga, Fernando. 2006. *MapuDungun: El habla Mapuche*. Santiago: Centro de Estudios Públicos.

INTERVIEWS BY AUTHOR

Antinao Varas, Clorinda [aka Clara Antinao Varas]. Santiago, Chile. July 20, 2012.

Antinao Varas, Clorinda [aka Clara Antinao Varas]. Santiago, Chile. December 26, 2015.

Avendaño Curaqueo, Elisa. Francisco Quereban Community, Commune of Padre de las Casas, Province of Cautín, Chile. December 20, 2023.

Curivil Paillavil, Ramón. Pedro Alonso Community, Oñoico, Commune of Saavedra, Province of Cautín, Chile. May 28, 2022.

Cuyanao, Jaime (Waikil). Via Zoom. July 8, 2024.

Gamboa, Luis. Via Zoom. July 10, 2024.

Huenchual Millaqueo, Margarita Elizabeth. Santiago, Chile. August 1, 2018.

Huinao, Graciela. Via Zoom. May 16, 2024.

Kakilpan, Francisco. Lican Ray, Commune of Villarrica, Province of Cautín, Chile. July 1, 2010.

Lienlaf, Leonel. Alepue, Commune of San José de la Mariquina, Province of Valdivia, Chile. January 5, 2015.

Mariman Quemenado, Pablo. Via Zoom. May 21, 2023.
Meulen Tranayado, Juan Atanacio. Osorno, Province of Osorno, Chile, December 21, 2015.
Millaleo, Ana. Via Zoom. June 17, 2024.
Navarro Caurepan, Arturo. Tralcao, Commune of San José de la Mariquina, Province of Valdivia, Chile. August 13, 2007.
Paillal Huechuqueo, José. Santiago, Chile. December 18, 2012.
Paillan Coñoepan, Elías. Via Zoom. May 31, 2023.

VIDEOS AND SOUND RECORDINGS

Aillapan Cayuleo, Lorenzo. 2008. *Wünül: Concierto de pájaros*. Documentary video, directed by Javiera Gallardo and Boris Muñoz. Delestero Realizadora.
Avendaño Curaqueo, Elisa. 2010. *Auktnkol ñi Vlkantun, hace eco mi música*. Documentary video produced by Elisa Avendaño Curaqueo in collaboration with Rosamel Millaman, Claudio Melillan, and Guido Brevis. Centro de Documentación Indígena, July 20, 2021, YouTube, 23 mins., 58 secs. https://www.youtube.com/watch?v=gdhTmydp1e8.
Avendaño Curaqueo, Elisa. 2023a. "Selinda" [or "Celinda"]. Music video posted by Orchard Enterprises for the Fundación Elisa Avendaño Curaqueo, May 26, 2023, YouTube, 4 mins., 42 secs. https://www.youtube.com/watch?v=lI2YDtJ2Zvg.
Avendaño Curaqueo, Elisa. 2023b. "Wilipag" [or "Wilipang"]. Music video posted by Orchard Enterprises for the Fundación Elisa Avendaño Curaqueo, May 26, 2023, YouTube, 5 mins., 10 secs. https://www.youtube.com/watch?v=LpoeAFocKNM.
El Indio Pije (Ernesto Ruiz), with La Sonora Trutruquera. 1975. *Qué pasa en la ruca: Show de cumbias*. LP. Santiago: Via Music.
Huichaqueo, Francisco. 2011. *Mencer Ñi Pewma*. Video. Santiago: Kinoko.
Wechekeche Ñi Trawün. 2005. *Wechekeche Ülkantun*. Album, CD. Santiago: Wechekeche Ñi Trawün.
Wechekeche Ñi Trawün. 2007a. *Kuifikeche Ñi Trawün*. Album, CD. Santiago: Wechekeche Ñi Trawün.
Wechekeche Ñi Trawün. 2007b. *Wechekeche Ka Kiñe*. Album, CD. Santiago: Wechekeche Ñi Trawün.
Wechekeche Ñi Trawün. 2009. *Wallmapuche/Wajmapuce: Ka Puel Kona*. Album, CD. Santiago: Wechekeche Ñi Trawün.
Wechekeche Ñi Trawün. 2015. "Mapudungufinge." Music video posted by Wetruwe Mapuche, January 16, 2015, YouTube, 6 mins., 41 secs. https://www.youtube.com/watch?v=h1-OzMy7hpY.
Wixage Anai. *Wixage Anai*. First broadcast, June 26, 1993. Digital copy, originally on cassette. Santiago: Centro de Comunicaciones Mapuce Jvfken Mapu.

INDEX

Kolibri 360 GmbH
Ferdinand-Jühlke-Straße 7
99095 Erfurt, Deutschland
produktsicherheit@kolibri360.de